Empirical Methods in Short-Term Climate Prediction

Empirical Methods in Short-Term Climate Prediction

Huug van den Dool
NOAA Climate Prediction Center

OXFORD
UNIVERSITY PRESS

OXFORD
UNIVERSITY PRESS

Great Clarendon Street, Oxford OX2 6DP

Oxford University Press is a department of the University of Oxford.
It furthers the University's objective of excellence in research, scholarship,
and education by publishing worldwide in

Oxford New York

Auckland Cape Town Dar es Salaam Hong Kong Karachi
Kuala Lumpur Madrid Melbourne Mexico City Nairobi
New Delhi Shanghai Taipei Toronto

With offices in

Argentina Austria Brazil Chile Czech Republic France Greece
Guatemala Hungary Italy Japan Poland Portugal Singapore
South Korea Switzerland Thailand Turkey Ukraine Vietnam

Oxford is a registered trade mark of Oxford University Press
in the UK and in certain other countries

Published in the United States
by Oxford University Press Inc., New York

British Library Cataloguing in Publication Data
Data available

Library of Congress Cataloging in Publication Data
Data available

Typeset by SPI Publisher Services, Pondicherry, India
Printed in Great Britain
on acid-free paper by
Biddles Ltd., King's Lynn, Norfolk

ISBN 0-19-920278-8   978-0-19-920278-2

1 3 5 7 9 10 8 6 4 2

Table of Contents

Dedication

To my late parents and teachers, Theresia E. M. Aengenent and Hugo M. van den Dool, Sr., as well as to my university professors in Utrecht, the late W. Bleeker and F. H. Schmidt. In naming them, I honor all of my teachers.

Foreword

A full century has passed since Albert Einstein announced his famous theory of relativity. Among the previously unobserved and unsuspected phenomena demanded by it was the bending of light rays that passed close to the Sun. Subsequent careful measurements confirmed that the theory was correct in this respect.

Meteorology is an example of a science where this sort of story cannot be told. When, for example, sounding balloons first revealed the presence of the stratosphere, also a century or so ago, no existing theoretical work had anticipated an extensive layer of the atmosphere in which the temperatures did not decrease with elevation. Somewhat more recently, when balloons began to reach higher elevations with some regularity, the discovery of the quasi-biennial oscillation in the equatorial stratosphere, where the wind blows generally from the east for about a year, and then from the west for about a year, came as a complete surprise. Even more recently the earliest reports of ozone depletion during the Antarctic spring were attributed to instrumental errors; no existing theory could account for them. In short, much of our continually growing knowledge of the nature of the atmosphere has been gained empirically.

Early attempts to predict the future weather were likewise empirical; even when Vilhelm Bjerknes proposed a century ago that the weather-forecasting problem was simply the problem of solving the dynamic equations that governed the atmosphere, he noted that no practical means of solving them was available. The earliest procedures were subjective, but, with the advent of electronic computers, it became possible to put into practice some of the systematic empirical methods that beforehand could at best be visualized. At the same time numerical weather prediction, based on the dynamic equations, began to flourish, but, despite the amazingly good, although by no means perfect, forecasts that are now produced one or two days ahead, the method still fails to outperform some much simpler empirical methods at ranges of a few weeks or months, presumably because of the chaotic nature of the atmosphere. Moreover, the procedures for parameterizing such processes as cumulus convection, even if based on dynamics, do not yield unique formulae, and the choice among several competitors tends to be empirical. It is thus entirely appropriate that a book devoted to empirical methods in atmospheric science should be forthcoming.

In this volume Dr. Huug van den Dool has assembled a collection of empirically derived objective forecasting procedures, including some that reveal positive skill at ranges where numerical weather prediction has yet to prove particularly effective. Some of these procedures are the products of his own research. Empirical wave propagation, to which he devotes the third chapter, is reminiscent of schemes that had been suggested many years ago; the difference is that as presently formulated it yields positive results. In a later chapter he discusses analogues—forecasting tools that were in use as early as World War II, but failed at that time to produce anything that was clearly superior to guesswork. In other chapters he describes procedures of more recent origin.

Dr. van den Dool looks at empirical procedures and numerical weather prediction as competing methods with a similar goal, and he speculates as to whether at some future date, at ranges of a few weeks or months, the latter method may outperform all others, thus rendering his book mainly of historical interest. This should be of no concern; it is likely to be the fate of any book that presents the current state of a developing field, and, in any event, if such a date arrives it will not do so tomorrow, and the practice of forecasting cannot meanwhile be suspended in anticipation of coming improvements. Assuming that such a date does arrive, I would guess that some empirical procedures will continue to prove superior at still longer ranges.

Edward N. Lorenz
Professor Emeritus of Meteorology
Massachusetts Institute of Technology
Cambridge, MA
March 2006

Preface

This book is based on my personal life-long trip through the field of climate prediction. Specifically, the writing of a book was inspired by my teaching various courses in recent years, most notably a graduate course with Eugenia Kalnay at the University of Maryland entitled "Atmospheric and Oceanographic Statistics", as well as the Climate Variations Course with several colleagues at the US National Weather Service (NWS). Similar inspiration was derived from outreach programs, where NWS personnel interact with user groups, as well as my recent presentations at conferences.

This book would have been impossible without the interaction with many colleagues over the years. To single out a few, Cor Schuurmans and Theo Opsteegh at the Royal Netherlands Meteorological Institute (KNMI) in the de Bilt, The Netherlands; Jerome Namias and John Horel at the Scripps Institution of Oceanography, La Jolla; William Klein, Ferdinand Baer and Ming Cai at the University of Maryland; Donald Gilman and David Rodenhuis at the Climate Analysis Center, and Eugenia Kalnay, Suranjana Saha, Jae Schemm, Jeff Anderson, Jin Huang at NCEP. I also want to acknowledge my co-forecasters at the Climate Prediction Center, Jim Wagner, Ed O'Lenic, Tony Barnston, Bob Livezey and David Unger. I must have spent months of my life collaborating, debating, and arguing with each of them.

Some of my colleagues may recognize their own work in specific parts of the text. This is especially true for Ming Cai, Peitao Peng, Jae Schemm, Suranjana Saha, David Unger and Jin Huang. The text of the manuscript improved a great deal as a result of careful reviews by Jeff Anderson, Ming Cai, Michelle L'Heureux, Åke Johansson, Tony Barnston, Ed O'Lenic and Cor Schuurmans. I also acknowledge the help of Jon Hoopingarner, David Unger, Suranjana Saha and Oxford University Press in preparing the figures.

I want to acknowledge NOAA management as represented by Vern Kousky, Jim Laver and Louis Uccellini, who not only allowed me to write a book in my official capacity, but also gave a lot of encouragement.

I especially thank Edward N. Lorenz, emeritus Professor at MIT, for his willingness to write a foreword and give a perspective on the relative roles of theory, modeling and empiricism.

Finally, I would like to acknowledge the loving support and encouragement I received from my wife, Suru Saha, and both our families in the writing of this book.

Motto: *There are two kinds of knowledge. First, there is the knowledge one learns from teachers along the way. Second, there is the knowledge that comes naturally by instinct or intuition, by ideas that come, seemingly, out of nowhere. It is the second kind of knowledge that I hope to impart to the reader of this book.*

Acronyms and notions

A	Above Normal Class (tercile)
AC	Anomaly Correlation
AMIP	Atmospheric Model Intercomparison Project. In practice a multi-decadal model run with observed interannually varying global SST as lower boundary conditions.
AMS	American Meteorological Society
AO	Arctic Oscillation, renamed Northern Annular Mode (counterpart for SH's Southern Annual Mode)
B	Below Normal (tercile)
CA	Constructed Analogue
CAC	Climate Analysis Center (forerunner of CPC)
CCA	Canonical Correlation Analysis
CDAS	Climate Data Assimilation System, real-time continuation of Reanalysis
CDC	Climate Diagnostics Center
CLIPER	A method based on Climatology and Persistence
CFS	Climate Forecast System used at NCEP
CPC	Climate Prediction Center, one of the nine centers in NCEP
cpdf	Conditional pdf
CV	Cross-Validation (in verification)
DEMETER	A European Seasonal Prediction Project
d.o.f.	Degrees of Freedom
dof	Degress of Freedom
E	Climatological Probability for the three-class system (1/3rd)
ECD	Empirical Correlation Distribution
ECMWF	European Center for Medium-range Weather Forecasting
e.d.o.f.	Effective Degrees of Freedom
ENSO	Name for the combination of El Nino and Southern Oscillation
El Nino	Oceanic Phenomenon: Occasional warming of tropical Pacific Ocean lasting months to a few years.
EMC	Environmental Modeling Center, one of the nine centers in NCEP
EOF	Empirical Orthogonal Function, also called Principal Component Analysis
EOT	Empirical Orthogonal Teleconnection
EOT2	Extension of EOT2 across two dat sets
EV	Explained Variance
EWP	Empirical Wave Propagation
EWP1	Empirical Wave Propagation using zonal harmonics
EWP2	Empirical Wave Propagation using global spherical harmonics
GCM	General Circulation Model

IRI	International Research Institute for Climate Prediction (Palisades, New York)
La Nina	Oceanic Phenomenon: Occasional cooling of tropical Pacific Ocean lasting months to a few years.
LIM	Linear Inverse Model
l.o.p.	Limit of Predictability
mb	Unit for pressure: the *millibar* (old unit; still allowed) equal to hectoPascal
MJO	Madden and Julian Oscillation
MRK	Markov Model
N	Near Normal Class
NA	Natural Analogues
NAO	North Atlantic Oscillation
NCEP	National Centers for Environmental Prediction, Washington DC (part of NWS).
NH	Northern Hemisphere
NINO34	Area in the Pacific Ocean from 5°S to 5°N and 170°W to 120°W. Usually the SST averaged over this area.
NINO3.4	Area in the Pacific Ocean from 5°S to 5°N and 170°W to 120°W. Usually the SST averaged over this area.
NOAA	National Oceanographic and Atmospheric Administration, US Dept of Commerce
NWP	Numerical Weather Prediction
NWS	National Weather Service in the US. One of the major components of NOAA
OCN	Optimal Climate Normal
PCA	Principal Component Analysis (same as EOF)
pdf	Probability density function
PER	Persistence (as forecast method)
PNA	Pacific North-American pattern
POP	Principal Oscillation Patterns
Reanalysis	Major international project to re-analyze weather maps up and down the atmopshere and the ocean retroactively from the 1940s forward with a constant method
rms	Root-mean-square
rmse	Root-mean-square error
SC	Squared covariance
SH	Southern Hemisphere
SO	Southern Oscillation, a seesaw of mass between the Indian and Pacific Oceans
SOI	Southern Oscillation Index
SST	Sea-Surface Temperature
STV	Space–Time Variance
SV	Spatial Variance
SVD	Singular Value Decomposition
QBO	Quasi-Biennial Oscillation
WMO	World Meteorological Organization
Z500	The height at which pressure is 500 mb

List of Plates

Plate 1: Display of teleconnection for seasonal (JFM) mean 500 mb height. Shown are the correlation between the base point (noted above the map) and all other grid points (maps) and the time series of 500 mb height anomaly (geopotential meters) at the base points. Contours every 0.2, starting contours $\pm$ 0.3. In both maps and time series the color red (blue) is used for positive (negative) values. Data source: NCEP Global Reanalysis. Period 1948–2005. Domain 20°N–90°N. On the left a pattern referred to as the North Atlantic Oscillation (NAO). On the right the Pacific North American pattern (PNA).

Plate 2: EV(i), the domain variance explained by single grid points in % of the total variance of seasonal (JFM) mean Z500 over 1948–2005, using Equation (4.3). In the upper left for raw data, in the upper right after removal of the first EOT mode, lower left after removal of the first two modes, etc. Contours every 4%. Values in excess of 4% lightly shaded, in excess of 12% dark shading. The time series shown are the residual height anomaly at the grid point that explains the most of the remaining domain integrated variance.

Plate 3: Display of four leading EOT for seasonal (JFM) mean 500 mb height. Shown are the regression coefficient between the height at the base point and the height at all other grid points (maps) and the time series of the residual 500 mb height anomaly (geopotential meters) at the base points. In the upper left for raw data, in the upper right after removal of the first EOT mode, lower left after removal of the first two modes, etc. Contours every 0.2, starting contours $\pm$ 0.1. Data source: NCEP Global Reanalysis. Period 1948–2005. Domain 20°N–90°N. (A light post-processing was applied, See Appendix I of Chapter 5).

Plate 4: Display of the four leading EOFs for seasonal (JFM) mean 500 mb height. Shown are the maps and the time series. A post-processing is applied, see Appendix 1, such that the physical units (gpm) are in the time series, and the maps have norm=1. Contours every 0.2, starting contours $\pm$ 0.1. Data source: NCEP Global Reanalysis. Period 1948–2005. Domain 20°N–90°N. The seed base point (e.g. 65°N,50°W) is mentioned because the iteration towards EOF (as described in Appendix 2) starts from the EOT.

50 and 60%. The color and the letters A or B indicate the shift in probability towards Above (A, T red, P green) or Below (B, T blue, P brown). White areas labeled EC have climatological probabilities for all three classes.

Plate 12:   The looks of a tool used to make the seasonal prediction. The tool is the CCA. The units are standard deviations multiplied by 10. Red (blue) values are positive (negative) anomalies. The size of the numeral indicates the level of a priori skill. At the locations, indicated by only the red or blue plus sign, the a priori skill is below the 0.3 correlation, and no forecast value is shown.

List of symbols

$\langle\ \rangle$	time mean
$\{\ \}$	a reference time mean (may not equal the time mean of data set under investigation)
$[\ ]$	a zonal or a space mean
$'$	departure from a time mean (or climatological mean)
$*$	departure from space mean
A	amplitude
A	verification attribute
$A(s)$	spatial map
a	regression coefficient
a	radius of the Earth
a, b	sine/cosine coefficients of harmonic expansion
$\mathbf{a}$	vector $\mathbf{a}$ in $\mathbf{Q}\,\mathbf{b} = \mathbf{a}$
$\mathbf{b}$	vector $\mathbf{b}$ in $\mathbf{Q}\,\mathbf{b} = \mathbf{a}$
b	regression coefficient
$\mathbf{B}$	general matrix
$B(s)$	spatial map
c, d	sine/cosine coefficients of harmonic expansion
c	intercept (regression)
c_{gx}, c_{gy}	zonal, meridonal component of group speed (m/s)
C	phase speed (m/s)
$\mathbf{C}$	time-lagged covariance matrix
c	a positive constant
d	basis function
e	basis function
$\mathbf{e}$	eigenvector (solution of $\mathbf{B}\,\mathbf{e} = \lambda\,\mathbf{e}$)
$\mathbf{E}$	matrix containing eigenvectors
$f(s, t)$	function of time and space, or a data set at n_s (n_t) points in space (time)
f	Coriolis parameter ($2\,\Omega \sin \phi$)
G	growth rate
$\mathbf{g}$	acceleration of gravity
$g(s, t)$	function of time and space, or a data set at n_s (n_t) points in space (time)
$h(s, t)$	function of time and space, or a data set at n_s (n_t) points in space (time)
F	Rossby radius of deformation

i, j	counter for position in time and space
k	mode number (for EOF, EOT, etc.)
K	multidimensional wavenumber
K	optimal number of years (for climatic normals)
L	sample size
$\mathscr{L}$	linear operator
M	size of a data set (e.g. number of years)
M	total number of orthogonal modes, $\leq$ the smaller of n_s and n_t
$M(s, t)$	complex mode
$\mathbf{M}$	diagonal matrix with elements μ_m
m	zonal wavenumber
m	mode number (for EOF, EOT, etc.)
m	month or season (1–12)
N	effective degrees of freedom
$N\mathscr{L}$	nonlinear operator
n	meridional wavenumber
n	total wavenumber (for associated spherical harmonics)
n_s	total number of gridpoints (or stations) in space
n_t	total number of time levels in a data set
P	constructed analogue prediction operator
P	precipitation
$\mathbf{Q}$	covariance matrix
$\mathbf{Q}^a$	alternative covariance matrix
q_{ij}	elements of $\mathbf{Q}$
$\mathbf{R}$	matrix of random forcing
S	spatial domain
s	space coordinate
SST	Sea-Surface Temperature
u, v	west–east, south–north wind components (m/s)
$u(t), v(t)$	time series
U	west to east background wind (m/s)
U	residual to be minimized
t	time (coordinate)
Δt	time increment
T	temperature
T	period
w	weighting factor
w	soil moisture
x	west–east coordinate
$\mathbf{x}$	state vector
y	south–north coordinate
Z	geopotential height (geopotential meter)
$Z500$	geopotential height of 500 mb height surface
$a(t)$	time series of expansion coefficients as in $f(s, t) = \Sigma a_m(t) e_m(s)$

a_j	weight multiplying an observed anomaly map
γ	threshold value
β	meridional derivative of the Coriolis parameter
$\beta(s)$	spatial pattern
$\beta(t)$	time series of expansion coefficients as in $f(s,t) = \Sigma\beta_m(t)e_m(s)$
ϵ	phase angle
ϵ	distance between two states of the atmosphere
ρ	damping factor
ρ	(auto)correlation
ρ_{ij}	shorthand for $\rho(s_i,s_j)$, the correlation between two positions in space
Ω	rotation of the Earth
τ	lead of the forecast; lag in (auto)correlation
λ, ϕ	longitude, latitude
λ	wavelength
λ	eigenvalue
Λ	matrix containing eigenvalues along the main diagonal and zero elsewhere
ψ	streamfunction. Under geostrophic assumption: $\psi = Z/f$
υ	probability
χ	velocity potential; any two-dimensional vector field $\mathbf{V}$, on a closed spherical domain, can be written $\mathbf{V} = (\partial\chi/\partial x - \partial\psi/\partial y, \partial\psi/\partial x + \partial\chi/\partial y)$
σ	singular value

1 Introduction

This is first and foremost a book about short-term climate *prediction*. The predictions we have in mind are for weather/climate elements, mainly temperature (T) and precipitation (P), at lead times longer than two weeks, beyond the realm of detailed Numerical Weather Prediction (NWP), i.e. predictions for the next month and the next seasons out to at most a few years. We call this *short-term climate* so as to distinguish it from long-term climate change which is not the main subject of this book. A few decades ago "short-term *climate* prediction" was known as "long-range weather prediction".[1] In order to understand short-term climate predictions, their skill and what they reveal about the atmosphere, ocean and land, several chapters are devoted to constructing prediction *methods*.

The approach taken is mainly *empirical*, which means literally that it is based in experience. We will use global data sets to represent the climate and weather humanity experienced (and measured!) in the past several decades. The idea is to use these existing data sets in order to construct prediction methods. In doing so we want to acknowledge that every measurement (with error bars) is a monument about the workings of Nature. We thought about using the word "statistical" instead of "empirical" in the title of the book. These two notions overlap, obviously, but we prefer the word "empirical" because we are driven more by intuition than by a desire to apply existing or developing new statistical theory.

While constructing prediction methods we want to discover to the greatest extent possible how the physical system works from observations. While not mentioned in the title, *diagnostics* of the physical system will thus be an important part of the book as well. We use a variety of classical tools to diagnose the geophysical system. Some of these tools have been developed further and/or old tools are applied in novel ways. We do not intend to cover all diagnostics methods, only those that relate closely to prediction.

---

[1] The distinction between *weather* and *climate* prediction is roughly along the lines of deterministic vs probabilistic. As long as forecasts are presented deterministically in the short range the term *weather prediction* is used. The larger uncertainty makes presentation of longer lead prediction probabilistic, more or less by necessity, hence climate prediction. More on this in Section 9.6

There will be an emphasis on *methods* used in operational prediction. It is quite difficult to gain a comprehensive idea from existing literature about methods used in operational short-term climate prediction. There are many articles on any one method in a research environment. Very little gets written about operational methods and activities. The emphasis on empiricism is also unusual. Many (review or research) articles have been written in the last two decades on seasonal prediction by the extension of NWP technology to climate (for instance, Shukla et al. 2000). Overview articles rarely described empirical methods, an exception being a brief overview by Hastenrath (2003). Empirical methods have not only survived, they have been further developed. This book should fill some of the gaps.

There are no particular geographical limitations in this book. Most methods discussed would potentially be applicable anywhere on the planet, even if the examples given are somewhat biased towards the experience in the United States. As much as possible we have included examples and illustrations for the often overlooked Southern Hemisphere. The difference between the hemispheres is striking and especially helpful when one wants to know how the system works. Nature has two experiments going on simultaneously in the two hemispheres with somewhat different settings.

Where we show illustrations, while we do refer to older literature (and give credit), new calculations were made in virtually all instances, using the 1948–present global NCEP/NCAR CDAS-Reanalysis (Kalnay et al. 1996; Kistler et al. 2001). This data set includes many variables (wind, pressure, temperature) at many levels in the atmosphere, but also global sea surface temperature (SST) and is kept up to date. Over land we use soil moisture (w) generated by running the Huang et al. (1996) model. SST and w are thought to be the source of some (if not most) of the memory in the geophysical system leading to short-term climate prediction skill. A large number of figures and verification results refer to 500 mb height, a traditional choice. This variable, half-way up in the atmosphere, has been the subject of extensive diagnostic study and forecast verification since upper air data became available.

In some figures data all the way through late 2005 were used. Indeed the author has always felt the confrontation of ideas with the real-time experience to be especially appealing.

Although we emphasize the empirical approach there will be plenty of reference to Numerical Weather Prediction (NWP) and its recent extensions into climate modeling. After all the goals are similar, so we will make comparisons and refer to concepts widely known by dynamicists.

Although we emphasize time-scales beyond two weeks, we will discuss application of empirical methods to time-scales shorter than two weeks where this is helpful in demonstrating how these methods work. This is especially true for understanding teleconnections. While teleconnections are usually mentioned in short-term climate prediction (like the effect of

ENSO on the extratropical winter mean state in the Northern Hemisphere), teleconnections owe their existence to processes operating on a day-by-day instantaneous time-scale. There is also the puzzling question why some methods, which have been retired for short-range weather prediction a long time ago, should still be competitive in the longer ranges.

There will be plenty of verification results in many of the chapters, but verification *methods* are not the subject of this book. Only traditional verification measures, anomaly correlation (AC), root-mean-square error (rmse), etc. are used. We occasionally mention a notion called cross-validation (CV) which applies to situations when retroactively made forecasts need to be verified. CV means that the verification datum is not used in any way when developing the forecast method, so as to ensure its independence as if it were future data (Michaelsen 1987). For recent developments in verification itself, including probability scores, the reader is referred to the book edited by Jolliffe and Stephenson (2003).

The seasonal forecast, regardless of method, has been around a long time. For instance the NWS in the US started experiments in 1958. But even before that, efforts were underway in many countries. Opinions differ as to whether the methods used at that time were an art or a science. More recently there have been several huge advances in this field that have made the approach more methodical. The first is the steady increase in the size of global data sets, punctuated by global Reanalysis in the mid-1990s which made the data set available to a wide audience. This obviously is the basis for empirical prediction. The second is the recognition of the global ocean as an external memory that could aid seasonal forecasts in the atmosphere. Special mention goes to the El Nino-Southern Oscillation phenomenon (ENSO), where redistribution of SST anomalies and tropical convection may impact areas very far away through teleconnections.[2] The third advance is the development of numerical methods for climate prediction. This is a natural extension of NWP to short-term climate prediction. The fourth major advance is the development of statistical-empirical methods, including an honest assessment of skill by retroactively made forecasts. In view of these recent advances it is perhaps timely to write this book.

The contents of this book are at the intersection of three scientific disciplines. Firstly there is geophysical fluid dynamics, which is the basis for much of modern meteorology and oceanography. Secondly, there is a large dose of statistics; this follows naturally from using large data sets. Thirdly there is applied mathematics because some of the methods are rooted in basic mathematical concepts like eigenanalysis, waves, etc.

While utility has driven most of the research in prediction methods, there is also the noble pursuit of knowledge. Not everything mentioned in this

---

[2] Before ENSO was a hot topic, Jerome Namias, grandfather of long-range weather prediction in the US, was a lone voice pushing the state of the ocean as predictor for the atmosphere.

book has proven to be practically useful. Whether it could be made useful is not always known ahead of time.

Another element of this book will be an attempt to simplify methods to their bare essentials. There is also a very practical side to the material since the author has been close to the operational forecast for many years.

This book abounds with 'Rock in the Pond' experiments. The proverbial experimental physicist who threw a rock in the pond had plenty to think about before he/she could explain the ripples emanating from the place of impact, reflections from the boundary, etc. Our typical experiment is to place an idealized disturbance of specified diameter at a certain location in a physical setting and wonder what will happen next.

The nine remaining chapters are organized as follows. In Chapter 2 we have collected some of the mathematics and statistics that underly the rest of the book and is specifically used. Chapter 3 presents a forecast method called Empirical Wave Propagation (EWP). Given a large data set, EWP can achieve a modest level of forecast skill with exceedingly simple means. Moreover EWP is a basis for understanding teleconnections as per wave propagation on the sphere. Chapter 4 deals with one-point teleconnections in the atmosphere. While teleconnections are not predictions as such, they describe the mechanism that makes, for instance, ENSO relevant to the mid-latitudes. A variant, empirical orthogonalized teleconnections (EOT), leads naturally to the subject of Chapter 5 which deals with empirical bi-orthogonal functions (EOFs), which are the cornerstone of a majority of modern empirical prediction methods. Use of EOFs for diagnostic purposes is also exceedingly common these days. Chapter 6 is a brief discussion of estimates of the degrees of freedom in the atmosphere—how many processes appear to be going on independently? As in Chapter 3 (EWP), Chapter 7 on analogues is an account of a method pioneered by the author. While natural analogues fail in most circumstances (of practical prediction interest) the constructed analogue (CA) is presented as a solution for the lack of data one would need for natural analogues to be a success. CA can be used for many different purposes; we discuss prediction of global SST, specification of fields of one variable from another, calculating (by empirical means) the fastest growing modes, etc. Chapter 8 gives a list plus discussion of nearly all methods used in modern short-term climate prediction plus examples of most methods. The list includes more than 90% of methods used operationally in the US. There is also an attempt to list, but with less detail, other methods that have been mentioned in the literature, and a discussion about the consolidation of a multitude of different forecasts. Chapter 9 is a look in the kitchen as to how the seasonal forecast is made in practice, including issues that relate to protocol, assumptions as to what users want and understand, managerial attitudes, etc. Chapter 10 is a wrap-up and conclusion including an attempt to explain why NWP, when applied to short-term climate prediction, is not necessarily better than a simple empirical method.

Most chapters can be read largely in isolation. Some reviewers suggested placing Chapter 3 (EWP) after Chapters 4 and 5. The reader can change the order of those chapters with only minimal problems. Experienced researchers may jump immediately to Chapter 8 since it comes closest to the contents promised in the title of the book.

With the appearance of a new book, one wonders which previous texts to compare to. Teleconnections and EOFs (our Chapters 4 and 5) have been treated in several other good textbooks over the past years. The books by Wilks (1995, 2005) and von Storch and Zwiers (1999) are obvious recent references on EOF in meteorology, while Jolliffe (2002) reaches across many disciplines. Peixoto and Oort (1992), while covering EOF and tele-connections as well, would be a recommended text for diagnostics (or General Circulation Statistics as it used to be called) in general. For most other chapters there are no recent textbooks.

The level of the book varies by chapter but is not basic, i.e. the accessa-bility for complete outsiders and undergraduates is limited. It is a book for graduate students, interested researchers and practitioners in short-term climate prediction. Chapter 9 and some of Chapter 8 are easier to read. There is little rigorous derivation. Rather, we apply dynamics, statistics and mathematics in intuitive ways, assuming the reader already knows the basics about regression, time series analysis, Rossby waves or solving a linear system of equations, etc. Phrased in terms of prerequisites the readers will thus benefit most if they are already familiar with the basics of atmospheric dynamics or oceanography, basic statistics, especially regression and spectral analysis, and linear algebra, especially eigenanaly-sis. It helps to know something about spectral prediction models used in weather prediction.

There are quite a few historical notes in this book. After all this is an old field and there are almost no previous texts. To the extent that NWP will outperform empirical methods at some point in the future the ironic reader may feel the whole book, because of its emphasis on empiricism, is part of writing history. Maybe so. However, assessments (usually no better than belief) about the importance of empirical methods for short-term climate prediction in the future differ. In Chapter 10 we critically review the scientific basis for expecting (or not) that numerical methods should outperform empirical methods in the future.

Long ago, in the nineteenth century say, it may have seemed to many that nature consists of a sum of exact cycles. This would suggest that observing for a long enough time, followed by discovery of all cycles and preferably a theoretical underpinning of each cycle, would eventually lead to perfect prediction out to infinity. The approach worked more or less for the motion of heavenly bodies, so why not for the atmosphere and ocean? Today we think of the atmosphere and ocean mainly as chaotic systems, sensitive to many details and with a finite prediction horizon. This has not lessened the need for observations. But the only periodic (and thus infinitely predictable) components in the system, those that are forced, are the daily and annual cycle (mainly heating) and the tides, mainly gravitational (thermal) in the ocean (atmosphere). Studies of predictability, i.e. assessing how good forecasts of the remaining anomalies could be theoretically, indicate a sobering perspective for short-term climate prediction. The reader studying prediction methods in this book or elsewhere should certainly be aware that even under ideal circumstance prediction skill may not be close to perfect. The main hope is to detect those coupled atmosphere–land–atmosphere components that have predictability time-scales longer than what the troposphere on its own is thought to possess (a few weeks).

2 Background on Orthogonal Functions and Covariance

The purpose of this chapter is to present some basic mathematics and statistics that will be used heavily in subsequent chapters. The organization of the material and the emphasis on some important details peculiar to the geophysical discipline should help the reader.

2.1 Orthogonal functions

Two functions f and g are defined to be orthogonal on a domain S if

$$\int_S f(s)\, g(s)\, ds = 0 \tag{2.1}$$

where s is space, one, two or more dimensions, and the integral is taken over S. Since we will work with data observed at discrete points and with large sets of orthogonal functions we redefine and extend a discrete version of (2.1) as

$$\sum_{s=1}^{n_s} e_k(s)e_m(s) = \begin{cases} 0 & \text{for } k \neq m \\ \text{positive} & \text{for } k = m \\ 1 \text{ for} & k = m; \text{ orthonormal.} \end{cases} \tag{2.2}$$

The summation is over $s = 1$ to n_s, the number of (observed) points in space. The $e_k(s)$ are basis functions, orthogonal to all $e_m(s), k \neq m$. The functions $e_k(s)$ are said to be orthonormal when the right-hand side in (2.2) is either zero or unity. For non-equal area data, such as on a latitude/longitude grid, or for irregular station data, the summation involves a weighting factor $w(s)$, i.e.

$$\sum_s w(s)\, e_k(s)\, e_m(s), \quad 1 \leq s \leq n_s$$

where $w(s)$ is related to the size of the area each data point represents. Except when noted otherwise, $w(s)$ will be left off throughout the book for simplicity. However this detail is not always trivial.

The $e_m(s)$ can be thought of as vectors consisting of n_s components. In that context the type of product in (2.2) is often referred to as inner or dot product $e_k \cdot e_m$.

A major convenience of a set of orthogonal functions $e_m(s)$ satisfying Equation (2.2) is functional representation of data, that is to say, for any discrete $f(s)$, say a map of mean sea level pressure on a grid, we can write:

$$f(s) = [f] + \sum_{m=1}^{M} \alpha_m \, e_m(s) \quad 1 \le s \le n_s \tag{2.3}$$

where $[f]$ is the spatial mean, α_m is the expansion coefficient, and M is at most $n_s - 1$. In the context of (2.3) it is clear why the $e_m(s)$ are called a basis. The equal sign in (2.3) only applies when the basis is complete.

For now, let's consider $[f]$ to be zero (or removed from the data as per $f^* = f - [f]$, then drop the *). We now define the Spatial Variance (SV) as:

$$SV = \sum_{s} f^2(s)/n_s \tag{2.4}$$

and note that SV, for orthonormal $e_m(s)$, can also be written as

$$SV = \sum_{m=1}^{M} \alpha^2_m \tag{2.5}$$

which establishes the link between variance in physical (2.4) and spectral space (2.5). The equals sign in (2.5) only applies when M is the required minimum value, which could be as high as $n_s - 1$. As per (2.5) each basis function "explains" a non-overlapping part of the variance, or (put another way) contributes an independent piece of information contained in $f(s)$. α^2_m is the classical Fourier power spectrum if the e's are sine/cosine on the domain. In that context (2.5) is known as Parceval's theorem, and counting the sine/cosine pair as one mode, M is at most $n_s/2$.

When e is known, then α_m can be easily calculated as:

$$\alpha_m = \sum_{s} f(s) \, e_m(s) \quad 1 \le m \le M \tag{2.6}$$

i.e. one finds the projection coefficients α_m by simply projecting the data $f(s)$ onto the e's on the points where they coincide. Equation (2.6) is only valid when the e's have unit length (orthonormal). If not, divide the right-hand side of (2.6) by $\sum e^2_m(s)$. Since orthogonal functions do not compete for the same variance, (2.6) can be evaluated for each m separately, and in any order. The ordering $m = 1$ to M is quite arbitrary. When sine/cosine is used, the low m values correspond to the largest spatial scales. But ordering by amplitude (or equivalently, explained variance) makes a lot of sense too.

If one were to truncate to N functions ($N<M$) the explained variance (EV) is given by (2.5), but summed only over $m = 1$ to N, so ordering orthogonal functions by EV is natural and advantageous for many purposes.

When the α_m and $e_m(s)$ are both known, the data in physical space can be retrieved via (2.3). When the α_m and $f(s)$ are known, a hypothetical situation, $e_m(s)$ cannot be retrieved in general.

It is necessary to reflect where the physical units reside in (2.3)–(2.6). If $f(s)$ is, say, pressure, in millibar (mb), the α_m assume the unit mb, and the SV has the unit mb^2. The $e_m(s)$ are dimensionless, and it is convenient for $e_m(s)$ to have unit length, such that the numerical values of α_m and SV make sense in physical units mb and mb^2, respectively.

The above was written for unspecified orthogonal functions. There is an infinity of orthogonal functions. Analytical orthogonal functions include the sin ms, cos ms pair (or cos $m(s-\epsilon)$) as the most famous of all; in this case (2.6) is known as the Fourier transform. Legendre polynomials, and spherical harmonics (a combination of sin/cos in the east–west direction and Legendre in the north–south direction) are also widely used in meteorology, starting with Baer and Platzman (1961). But the list includes Bessel, Hermite, Chebyshev, Laguerre functions, etc. Why prefer one function over the other? There are issues of taste, preference, accuracy, theory, scaling, tradition, ... convenience. We mention here specifically the issue of "efficiency". For practical reasons one may have to truncate, in (2.3), to much less than M. If only N orthogonal functions are allowed ($N<M$) it matters which orthogonal functions will explain the most variance. The remainder is relegated to unresolved scales, unexplained variance and truncation error. (A different type of efficiency has to do with the speed by which transforms like (2.3) and (2.6), can be executed on a computer.)

One can easily imagine non-analytical orthogonal functions. Examples include zeros at all points in space except one; this makes for a set of orthogonal functions equal to n_s. An advantage of analytical functions is that there is theory and a wealth of information. Moreover, analytical functions suggest values and meaning in between the data points. This makes differentiation and interpolation easy. Nevertheless, ever since Lorenz (1956), non-analytical empirical orthogonal functions (no more than a set of numbers on a grid) are highly popular in meteorology as a device to "let the data speak". Moreover, these empirical orthogonal functions (EOF) are the most efficient in explaining variance for a given data set.

We have written the above, (2.1)–(2.6), for space s. One can trivially replace space s by time $t(1 \leq t \leq n_t)$ and keep the exact same equations showing t instead of s. However, if one has a space–time data set, $f(s, t)$, as one typically does, the situation becomes quickly more involved. For instance, choosing orthogonal functions, as before, in space, (2.3) can be written as:

$$f(s,t) = [f(s,t)] + \sum_{m=1}^{M} \alpha_m(t) \, e_m(s) \quad 1 \le s \le n_s, \, 1 \le t \le n_t \qquad (2.7)$$

where the projection coefficients and the space mean are a function of time; While choosing orthogonal functions in time would lead to:

$$f(s,t) = \langle f(s,t) \rangle + \sum_{m=1}^{M} \alpha_m(s) \, e_m(t) \quad 1 \le s \le n_s, \, 1 \le t \le n_t \qquad (2.7a)$$

where the time mean, and the projection coefficients are a function of space. Time mean is denoted by $\langle \ \rangle$.

Equation(2.7) and (2.7a) lead, in general, to drastically different looks of the same data set, sometimes referred to as T-mode and S-mode analysis. There is, however, one unique set of functions for which this space–time ambiguity can be removed: EOFs, i.e $\alpha_m(t)$ in (2.7) is the same as $e_m(t)$ in (2.7a), and $\alpha_m(s)$ in (2.7a) is the same as $e_m(s)$ in (2.7). We phrase this as follows: For EOFs one can reverse (interchange) the roles of time and space. This does, however, require a careful treatment of the space–time mean, see Section 2.3.

2.2   Correlation and covariance

Here we discuss elementary statistics in one dimension first (time) and use, as a not-so-arbitrary example of two times series, $D(t)$, the seasonal mean pressure at Darwin in Australia near a center of action of a phenomenon called ENSO, and seasonal mean temperature $T(t)$ at some far away location in mid-latitude: $T(t)$, $1 \le t \le n_t$, where t is a year index, 1948–2005 say; $n_t = 58$). One can define the time mean of D as

$$\langle D \rangle = \sum_{t=1}^{n_t} D(t) \, / \, n_t, \qquad (2.8)$$

and T similarly has a time average $\langle T \rangle$. We now formulate departures from the mean, often called anomalies:

$$\left. \begin{array}{l} D'(t) = D(t) - \langle D \rangle \\ T'(t) = T(t) - \langle T \rangle \end{array} \right\} \quad \text{for all } t. \qquad (2.9)$$

The covariance between D and T is given by

$$\text{cov}_{DT} = \sum_{t} D'(t) T'(t) / n_t. \qquad (2.10)$$

The physical units of covariance in this example are (mb $^\circ$C). The variance is given by

$$\text{var}_D = \sum D'(t)D'(t) / n_t, \tag{2.11}$$

and similarly for var_T. The standard deviation is $\text{sd}_D = \sqrt{\text{var}_D}$, and the correlation between D and T is:

$$\rho = \text{cov}_{DT}/(\text{sd}_D \cdot \text{sd}_T). \tag{2.12}$$

The correlation is a non-dimensional quantity, $-1 \le \rho \le 1$.

Calling D the predictor, and T the predictand, there is a regression line $T'_{\text{fcst}} = bD'$ which, over $t = 1, n_t$, explains $\rho^2\%$ of the variance in T. The subscript 'fcst' designates a forecast for T given D. The regression coefficient b is given by

$$b = \rho \, \text{sd}_T/\text{sd}_D. \tag{2.13}$$

The correlation has been used widely in teleconnection studies to gauge relationships or "connections" between far away points. The suggestion of a predictive capability is more explicit when $D(t)$ and $T(t)$, while both time series of length n_t, are offset in time, D leading T. If D and T are the same variable at the same location, but offset in time, the above describes the first steps of an autoregressive forecast system. Note also that the correlation is used frequently for verification of forecasts against observations.

In many texts the route to (2.12) is taken via "standardized" variables, i.e. using (2.8), (2.9) and (2.11) we define:

$$D''(t) = (D(t) - \langle D \rangle) / \text{sd}_D \tag{2.9a}$$

which have no physical units. Given these standardized anomalies, correlation and covariance become the same and are given by

$$\rho = \text{cov}_{DT} = \sum_t D''(t)T''(t) / n_t. \tag{2.12a}$$

So, the correlation does not change with respect to (2.12), but the covariance does change relative to (2.10) and loses its physical units.

It should be trivial to replace time t by space s in (2.8)–(2.12) and define covariance or correlation in space in analogous fashion. Extending to both space and time, and using general notation, we have a data set $f(s, t)$ with a mean removed. The covariance in time between two points s_i and s_j is given by

$$q_{ij} = \sum_t f(s_i, t)f(s_j, t) / n_t \tag{2.14}$$

while the covariance in space between two times t_i and t_j is given by:

$$q^a{}_{ij} = \sum_s f(s, t_i) f(s, t_j) / n_s, \tag{2.14a}[1]$$

---

[1] When a non-equal area grid is used the expression has to be adjusted in a calculation as follows:

$$q^a{}_{ij} = \sum w(s) f(s,t_i) f(s,t_j) / W \tag{2.14a}$$

where $w(s)$ represents the size of the area for each data point in space, and W is the sum (in space) of all $w(s)$. On the common lat-lon grid the weight is cos(latitude).

where q_{ij} and $q^a{}_{ij}$ are the elements of the two renditions of the all-important covariance matrices $\mathbf{Q}$ and $\mathbf{Q}^a$; the superscript "a" stands for alternative. The $n_s \times n_s$ matrix $\mathbf{Q}$ measures "teleconnection" between any two points s_i and s_j while the $n_t \times n_t$ matrix $\mathbf{Q}^a$ measures the similarity of two maps at any two times t_i and t_j, a measure of analogy. These two features (teleconnection and analogy) are totally unrelated at first sight, but under the right definitions the eigenvalues of $\mathbf{Q}$ and $\mathbf{Q}^a$ are actually the same, such that the role of space and time can be thought of as reversible. One issue to be particularly careful about is the mean value that is removed from $f(s, t)$. This has strong repercussions for everything we have said to far in Chapter 2.

2.3   Issues about removal of "the mean"

An important but murky issue is that of forming anomalies. The general idea is one of splitting a datum into a part that is easy to know (some mean value we are supposed to know), and the remainder, or anomaly, which deals with variability around that mean and is a worthy target for prediction efforts. All attention is subsequently given to the anomaly. Should it be $f'(t) = f(t) - \langle f \rangle$ as in Equation(2.9) or $f'(t) = f(t) - \{f\}$, where $\{f\}$ is a reference value, not necessarily the sample time mean. This is a matter of judgement. Examples where this question arises:

(a) when the mean is known theoretically (there may be no need to calculate a flawed mean from a limited sample);
(b) when forecasts are made of the type: "warmer than normal", with respect to a normal which is based on past data by necessity at the time of issuance of the forecast;
(c) The widely used anomaly correlation in verification, see the appendix of Chapter 2;
(d) In EOF calculations (a somewhat hidden problem);
(e) When the climatology is smoothed across calendar months, resulting in non-zero time mean anomalies at certain times of the year.

While no absolute truth and guidelines exist we here take the point of view that removal of a reference value, acting as an approximate time mean, is often the right course of action.

The removal of a space mean is not recommended and $\sum_s f'(s) \neq 0$ and $\sum_t f'(t) \neq 0$ are acceptable. Removal of a calculated space mean is problematic. On planet Earth, with its widely varying climate, removing a space mean first makes little sense as it creates, for example, anomalies warmer (colder) than average in equatorial (polar) areas.

Given a space–time data set $f(s, t)$ we will thus follow this practice:

(1) Remove at each point in space a reference value $\{f(s)\}$, i.e. form anomalies as follows:

$$f'(s, t) = f(s, t) - \{f(s)\}. \tag{2.15}$$

In some cases and examples the $\{\ \}$ reference will be the sample time mean, such that anomalies do sum up to zero over time and $f'(t)$ is strictly centered. But we do not impose such a requirement.

(2) We do NOT remove the spatial mean of f'.

Under this practice we evaluate (2.14) and (2.14a). We furthermore note that under the above working definition the space–time variance is given by

$$\text{STV} = \sum_{s,t} f'^2(s,t)/(n_s\, n_t). \tag{2.16}$$

Exactly the same total variance is to be divided among orthogonal functions (EOF or otherwise) calculated from either $\mathbf{Q}$ or $\mathbf{Q}^a$. We acknowledge that some authors would take an additional space mean out when working with $\mathbf{Q}^a$ (because they feel they should require the sum of the anomalies in space to be zero). This, however, modifies the STV, and all information derived from $\mathbf{Q}^a$ would change. We do not recommend taking the space mean out. Especially on small domains, taking out the space mean of f' takes away much of the signal of interest. We thus work with anomalies that do not necessarily sum up to exactly zero in either the time or space domain. This also means that variance and standard deviation, as defined in (2.11), (2.12) and (2.16), are augmented by the (usually small) offset from zero mean. When calculating EOFs the domain means get absorbed into one or more modes.

2.4 Concluding remarks

We have approached the representation of a data set $f(s, t) = [f(s, t)] + \sum \alpha_m(t)\, e_m(s)$ both by classical mathematical analysis theory and by basic statistical concepts that will allow calculation of orthogonal functions from a data set (mainly in Chapter 5). It may be a good idea to reflect on the commonality of Sections 2.1 and 2.2 and the juxtaposition of terminology. For example, the inner product used to determine orthogonality (2.1) is the same as the measure for covariance (or correlation) in (2.10). Indeed a zero correlation is a sure sign of two orthogonal time series or two orthogonal maps. In both Sections 2.1 and 2.2 we mentioned the notion of explained variance (EV), once for orthogonal functions, once for regression. We invited the reader to follow the physical units, and numerically the numbers should make physical sense when the basis is orthonormal. This task becomes difficult because the use of EOFs allows basis functions to be orthogonal in time and space simultaneously—both are basis function and projection coefficients all at the same time. We emphasized the "reversibility" of time and space in some of the calculations. We finally spent some paragraphs on removing a mean, which, while seemingly a detail, can cause large differences in interpretation.

Appendix: The anomaly correlation

One of the most famous correlations in meteorology is the anomaly correlation used for verification. It has been in use at least since Miyakoda et al (1972). Imagine we have, as a function of latitude and longitude, a 500 mb height field $Z(s)$. The field is given on a grid, cosine weights not shown. How do we correlate two 500 mb height maps, like for instance Z_{fcst} and Z_0, a set of paired forecast/observed maps, a challenge faced each day by operational weather forecast centers? The core issue is forming anomalies or splitting Z into a component we are supposed to know (no reward for forecasting it right) and the much tougher remainder. A definitely 'wrong way'[2] would be to form anomalies by $Z^*(s) = Z(s) - [Z]$ where $[Z] = Z(s)/n_s$, a space mean. Removal of a calculated space mean is problematic. On planet Earth with its widely varying climate, removing a space mean makes little sense as it creates, for example, anomalies warmer/colder than average in equatorial/polar areas. Even a terrible forecast would have a high anomaly correlation.

A "better way"[2] abbreviate is to form anomalies via $Z'(s) = Z(s) - Z_{climo}(s)$ and likewise $Z'_{fcst}(s) = Z_{fcst}(s) - Z_{climo}(s)$ where $Z_{climo}(s)$ is based on a long multi-year observed data set $Z(\lambda, \phi,$ pressure level, day of the year, hour of the day, etc.). The anomaly correlation is then given by

$$AC = \frac{\sum Z'_{fcst}(s)Z'_0(s)/n_s}{[\sum Z'_{fcst}(s)Z'_{fcst}(s)/n_s \cdot \sum Z'_0(s)Z'_0(s)/n_s]^{1/2}} \qquad (2.17)$$

where summation is over space: $-1 \le AC \le 1$, cosine weights not shown.

Among the debatable issues: should $\sum_s Z'(s)$ be (made) zero? There is no reason to do that, especially in verification, but some people feel that way. The removal of the space mean removes a potentially important component of the forecast from the verification process. This is especially true on small domains that may be dominated by anomalies of one sign.

---

[2] There is no absolutely right or wrong in these issues. In a 2D homogeneous turbulence experiment time and space means would be expected to be the same, so taking out the space mean may be quite acceptable.

Note that $-1 \le AC \le 1$, like a regular correlation, even though the spatial means are not exactly zero and the two terms in the denominator are augmented versions of the classical notion variance.

Equation(2.17) is for a single pair of maps. When we have a large set of paired maps we sum the three terms in (2.17) over the whole set, then execute the multiplication and division.

3 Empirical Wave Propagation

The purpose of this chapter is to demonstrate that, given a long data set of global extent, one can design a simple forecast method called Empirical Wave Propagation (EWP), which has modest forecast skill and allows us to explore aspects of atmospheric dynamics *empirically*, most notably aspects that help to explain mechanisms of teleconnection. The highlight of this chapter are dispersion experiments where we ask the question what happens to an isolated source at $t = 0$? Even though Nature has never done such an experiment, we will address this question empirically. In case the reader does not need/want to know the technical details of deriving wavespeeds he/she can skip to page 22 (EWP diagnostics sct 3.2) of this chapter. We will also discuss the skill of one-day EWP forecasts, in comparison to skill controls like "persistence", as a function of season, hemisphere, level and variable. While short-range (1 day) forecasts are certainly not the topic of this book, we note that the short-term wave propagation features described here do nourish and maintain the teleconnection patterns thought to be important for longer range forecasts.

EWP uses either zonal harmonic waves (sin/cos pairs) along each latitude circle separately, or global domain spherical harmonics (see Parkinson and Washington (1986) for the basics on spherical harmonics). The orthogonal functions used here are thus analytical. The atmosphere is to first order rotation-symmetric and obviously periodic in the east–west direction, which makes the zonal Fourier transform a natural. Moreover, many weather systems, wave-like in the upper levels, are seen to move from west to east (east to west) in the mid-latitudes (tropics), so a decomposition in sin/cos functions should inform us about phase propagation and energy dispersion on the sphere. For any initial time we decompose the state of the atmosphere into harmonic waves. If we knew the wave speed, and made an assumption about the future amplitude, we could make forecasts by analytical means.[1] But how do we know the phase speed? One way to proceed, with data

---

[1] There is an implicit assumption that waves of different wavenumber travel independently, i.e. no nonlinear interaction.

alone, is to calculate from a large data set the climatological speeds of anomaly[2] waves. This is where the empirical aspects come in. Phase speed estimates can be made via a technique called phase shifting.

3.1   Data and EWP method

3.1.1   Data treatment

Consider a data set of, for example, 500 mb height analyses (treated as "observed"), once daily at 0Z, on a 2.5° × 2.5° lat/lon grid, denoted as $Z(\lambda, \varphi, t, \text{year})$, where λ, φ are longitude and latitude. Choosing just a small window in the annual cycle (± 15 days) around January 15, we can combine all January days during 1979–1995 into one single data set and have $t = 1, 31$. We now form anomalies by:

$$Z'(\lambda, \varphi, t, \text{year}) = Z(\lambda, \varphi, t, \text{year}) - Z_{\text{climo}}(\lambda, \varphi, t)$$

where $Z_{\text{climo}}(\lambda, \varphi, t)$ is based on a long multi-year data set $Z(\lambda, \varphi, p\text{-level}, \text{day of the year, hour of the day, etc.})$. See Schemm et al. (1997) for details on how such climatologies are prepared. Global Reanalysis (Kalnay et al. 1996) and CDAS, its continuation in real time (Kistler et al. 2001), allow us to choose any subperiod during 1948–present. We now select data along just one latitude circle at a time, a periodic domain: $Z'(\lambda, 50N, t, \text{year})$. We further simplify notation to $Z'(\lambda, t)$.

As in Equation(2.6) we project anomaly data Z' onto the sin / cos orthogonal pair for each t. This yields two coefficients (a and b) , or, alternatively, an amplitude (A) and a phase (ϵ) for each m, $m = 0$ to 72, i.e.

$$Z'(\lambda, t) = A_0(t) + \sum_m a_m(t) \cos mx + b_m(t) \sin mx$$

$$= A_0(t) + \sum_m A_m(t) \cos m(x - \epsilon_m(t)), \tag{3.1}$$

where $x = 2\pi\lambda/360$, and $\lambda = 0, 2.5, 5, \ldots, 357.5$. A_0 is the zonal mean of Z', sometimes referred to as wavenumber 0; ϵ_m is the phase angle. Equation (3.1) is a classical Fourier transform or harmonic analysis of $Z'(\lambda, t)$. Recall, Z', A_m and ϵ_m are all functions of time.

3.1.2   Amplitude

Essentially, in view of (3.1), in order to forecast $Z'(\lambda, t + 1)$ given $Z'(\lambda, t)$, we seek information about the amplitude A_m and the phase ϵ_m at $t + 1$. Splitting up the forecast problem explicitly into these two aspects (A, ϵ) is not all that common, but leads to special insights. The real forecast skill

---

[2] Anomaly is defined as a departure from a climatological mean.

resides in the propagation aspect, while the skill related to the amplitude, in data studies, is usually just damping. Nevertheless we can learn from studying first the amplitude. In view of Parceval's theorem. Equation (2.16), we can write the space–time variance (STV) as:

$$STV = \sum_t \sum_s Z^{\prime^2}/(n_s n_t) = 1/2 \sum_t \sum_m A_m^2$$
$$= 1/2 \sum_m (\langle A_m \rangle^2 + \langle A_m^{\prime 2} \rangle), \qquad (3.2)$$

where $\langle \; \rangle$ is the time mean, and $A' = A - \langle A \rangle$.

From numerous calculations with many variables we find that about 75% or more of the variance in the atmosphere is associated with $\langle A_m \rangle^2$, i.e. the observed variability can be thought of as anomaly waves with amplitude fixed at their climatological value, $\langle A_m \rangle$, residing at some phase. (An example of this calculation is forthcoming in the discussion of Table 3.1.) The remainder, $\langle A'_m{}^2 \rangle$, due to amplitude variations is 25% or less, depending on variable. This certainly creates, by and large, the impression of stable waves, and therefore the prediction as one of primarily the phasing of waves. Striking "development" localized in space (such as a suddenly growing cyclone) has to be mainly one of constructive interference, not one of periodic sin/cosine wave amplitude development. This point of view is in agreement with Farrell (1984) who was one of the first to question whether "modal" (i.e. sin/cos) instability is the cause of mid-latitude cyclone development, a view that had been held since the 1940s.

Please note the second layer of climatology in Equation(3.2). We had already removed Z_{climo} from Z, then defined the climatology of the amplitude of anomaly waves (a second moment). In contrast to $\langle a_m \rangle$ and $\langle b_m \rangle$, $\langle A_m \rangle$ is not zero because amplitude is derived from a squared quantity (i.e. note that Equation(3.2) does not refer to the amplitude of climatological mean waves in Z).

Now, take a single anomaly wave $A_m \cos m(x-\epsilon_m)$. The question is: will this wave move east or west and by how much per unit time? The question cannot be trivially answered by studying data because (a) the speed varies greatly from day to day, and (b) ambiguities arise when the wave moves more than 180° to the east (or west), relative to its own wavelength. To lessen these problems we use a "phase shifting" technique.

3.1.3 Phase shifting

Following Van den Dool and Qin (1996), and dropping the m index in A and ϵ for simplicity, we consider a single wave m and write:

At time t: $A \cos m(x - \epsilon) = a \cos mx + b \sin mx$ $\qquad (3.3)$

At time $t+1$: $A^{+1} \cos m(x - \epsilon^{+1}) = a^{+1} \cos mx + b^{+1} \sin mx.$ $\qquad (3.3a)$

Now move the crest of the wave at time t to a reference longitude (Greenwich for instance); this is done by phase shifting over $+\epsilon$. Move the wave on the next time level $(t + 1)$ over the same angle ϵ; this maintains the relative positioning of the waves at successive days, but in a new framework. Phase shifting yields:

$$\text{At time } t: \quad A\cos\ m(x) = A\cos\ mx + 0\sin\ mx \tag{3.4}$$

$$\text{At time } t + 1: \ A^{+1}\cos\ m(x - (\epsilon^{+1} - \epsilon)) = c^{+1}\cos\ mx + d^{+1}\sin\ mx \tag{3.4a}$$

where $c^{+1} = a^{+1}\cos m\epsilon + b^{+1}\sin m\epsilon$ and $d^{+1} = b^{+1}\cos m\epsilon - a^{+1}\sin m\ \epsilon$.

The phase shifting is done for all pairs $t/(t+1)$ (all 510 pairs for, say, January 1979–1995 for instance) and ϵ is always the phase angle on the leading day. The right-hand side coefficients in (3.4) and (3.4a), A, c^{+1} and d^{+1}, are a function of time, with time means $\langle A\rangle$, $\langle c^{+1}\rangle$ and $\langle d^{+1}\rangle$. The time mean of coefficients a and b (and a^{+1} and b^{+1}) would be very nearly zero. All variables we have introduced in (3.3) and (3.4) can be evaluated from the data.

Amplitudes of the time averaged phase shifted (subscript ps) wave m are given by:

$$t: \quad A_{ps} = \langle A\rangle \tag{3.5a}$$

$$t + 1: \ A_{ps}{}^{+1} = \sqrt{(\langle c^{+1}\rangle^2 + \langle d^{+1}\rangle^2)}. \tag{3.5b}$$

Phase angles of the time averaged phase shifted wave are given by:

$$t: \quad \epsilon_{ps} = 0 \tag{3.5c}$$

$$t + 1: \ \epsilon_{ps}{}^{+1} = \arctan(\langle d^{+1}\rangle/\langle c^{+1}\rangle). \tag{3.5d}$$

The resulting A_{ps} and ϵ_{ps} can be generated for each m. The amplitude at time t is not changed by the phase shifting: A_{ps} is the same as $\langle A_m\rangle$ in (3.2).

If the wave at $t+1$ were in a random phase relative to the wave at t, $\langle c^{+1}\rangle$ and $\langle d^{+1}\rangle$ would be zero and hence A_{ps}^{+1} would be zero. The ratio A_{ps}^{+1}/A_{ps} thus tells us the degree of non-randomness in $(\epsilon^{+1} - \epsilon)$ or the steadiness in propagation for the given time increment.

The phase shifting technique is helpful mainly because it postpones dealing with the ambiguity about displacement larger than $\pm 180°$, until after the time averaging of c^{+1} and d^{+1}. On many individual days with either high phase speed and/or low amplitude waves (or too large time increment) the ambiguity is difficult to deal with.

EWP is related to time spectral analysis but uses only short-time increment lagged data to determine wave speeds under quasi-linear conditions.

3.1.4 Mean propagation

The phase speed C (in m/s) can be obtained from ϵ_p^{+1} (in (3.5d) in radians) as

$$C(\phi, m) = \epsilon_{ps}^{+1}(m, \phi).637\,5000.\cos(\phi)/86\,400/m \tag{3.6}$$

where the constants are the radius of the Earth and number of seconds per day (since we used data once daily). When using spherical harmonics instead of sin/cos only the speed at the equator needs to be reported. In all cases $A_{ps}^{+1} < A_{ps}$, i.e. using this method, anomaly wave amplitudes are always damped. Some degree of damping is typical for statistical methods. All waves appear stable. Damping is small (large) for long (short) waves. We have found that a one-day time increment works very well. For larger time spacing (2–10 days), the damping increases quickly.

3.1.5   EWP forecast method

To apply EWP as a forecast method it is enough to know ϵ_{ps}^{+1} as a function of wavenumber and latitude for a given time of year, i.e. on independent data we decompose the anomaly height field at $t = 0$ into waves, using (2.6), then move each wave by $\epsilon_{ps}^{+1}(m, \phi)$ from (3.5d), then use (3.1) to arrive at a forecast in physical space for $t = 1$. The very same forecast method could also be applied if $\epsilon_{ps}^{+1}(m, \phi)$ were known from theory, as is the case for a simple model (barotropic) in a simple basic state (like $U(\phi) = U_{eq} \cos \phi$, a state called super-rotation). EWP is an analytical prediction method, and the word empirical applies only to the source of information that yields ϵ_{ps}^{+1}.

In the above we presented and derived the EWP forecast method along intuitive lines. In Appendix 1, we also present a formal derivation based on rmse minimization with very nearly the same result.

3.2 EWP diagnostics

Table 3.1 serves as an example of the diagnostic aspects of EWP. We analyzed 20 years of 500 mb data for January 1968–1987. Information is given here for 50°N, for selected zonal wavenumbers $m = 0,1,3,5,7,9,11$. For quick comprehension all numbers are rounded off to the nearest integer.

The time mean amplitude $\langle A_m \rangle$ is given in the first line, and the fraction of variance represented by $\langle A_m \rangle$ in the second line. The long waves have large amplitude, nearly constant for $m=1$, to 4 at 70–75 geopotential meter (gpm) while amplitude drops off sharply with m beyond wavenumber 4. As shown in the second line, the time mean amplitude of the anomaly waves represents 75–80% of the variance (except in the zonal mean ($m=0$) where the percentage is only 63%). The phase propagation, in degrees relative to own wavelength, denoted $°\lambda$, is given in the third line and the conversion to speed in m/s in the fourth line. Long waves travel westward $(-)$, and short waves eastward $(+)$, in good qualitative agreement with the theoretical Rossby equation for mid-latitudes (Holton 1979, p.167; see Appendix 2 for more details), which reads $C = U - \beta/K^2$, where C is the phase speed, U is the background windspeed, β is the meridional derivative of the Coriolis parameter, and K is the wavenumber (if only the zonal wavenumber is considered K relates to m as $K = 2\pi m/L$, where L is the length of the latitude circle.) The short wave speed (large K or m) is nearly constant with m at 10 m/s and no ambiguities arise for the wavenumbers shown, the largest displacement shown being 136° or less than half the wavelength, even for $m=11$. The displacement in degrees depends obviously on the time increment (Δt), chosen here as one day, but the speed C in m/s depends barely on Δt as long as Δt is small. As can be judged from the fifth line, the phase propagation is rather steady (large A_{ps}^{+1}/A_{ps}) for the long waves, but is increasingly more variable and harder to determine for the short waves.

The fact that wave speed depends on m, or the wavelength, is called dispersion and leads to most interesting consequences described later on.

Table 3.1. Tabulation of amplitude, % variance, phase angle propagation, phase speed and amplitude ratio, for selected zonal wavenumbers of daily 500 mb height anomalies in January for 50°N. The time increment is 24 hours. Period = 1969–1987. ND = Not Defined.

>		Zonal wave number m						units	Reference
	0	1	3	5	7	9	11		
$\langle A_m \rangle$ or A_{ps}	26	73	73	57	35	21	13	gpm	Eq (3.5a)
$\langle A_m \rangle^2/\langle A_m^2 \rangle$	63	80	79	82	80	77	77	%	(3.2)
ϵ_{ps}^{+1}	ND	−3	3	31	71	108	136	$°\lambda$	(3.5d)
C	ND	−3	1	5	8	10	10	m/s	(3.6)
A_{ps}^{+1}/A_{ps}	89	90	88	79	65	44	32	%	(3.5a/b)

Table 3.2. The same as Table 3.1 but now 50°S. (Index m not shown)

>		0	1	3	5	7	9	11	units	Reference
	Zonal wavenumber									
$\langle A \rangle$ or A_{ps}		25	58	60	59	31	18	10	gpm	(3.5a)
$\langle A \rangle^2 / \langle A^2 \rangle$		68	79	79	77	77	79	80	%	(3.2)
ϵ_{ps}^{+1}		ND	−4	7	39	93	147	−167*	°λ	(3.5d)
C		ND	−3	2	6	11	14	−13**	m/s	(3.6)
A_{ps}^{+1} / A_{ps}		92	86	84	89	77	60	38	%	(3.5a/b)

* This value may be interpreted as +193.
** This value may be interpreted as +15m/s.

Table 3.2 is the same as Table 3.1, but now for the SH along 50°S. By and large the results are the same in the mid-latitudes of the two hemispheres in January, so we just point out a few salient differences.

Firstly, the motion of the mobile shorter waves in the range $m = 5$ to 10 in the SH is more steady; compare $A_{ps}{}^{+1}/A_{ps}$ in the fifth lines in Tables 3.1 and 3.2. This will turn out to be a great help in making EWP forecasts for the SH, where behavior is more wave-like, regular and less turbulent (Salby 1982). The second point of difference to note is that the value of ϵ_{ps}^{+1} is apparently greater than 180° already for wavenumber 11. This is because short-wave eastward speeds are larger in the SH than in the NH (January, 500 mb) and zonal wave 11 travels more than its own half-wavelength in one day in the SH. As indicated in the footnote of Table 3.2 we feel confident in resolving this ambiguity, and therefore substitute the appropriate positive numbers, but in general such ambiguities are problematic. It actually depends on the application whether the ambiguity needs to be addressed at all. For a diagnostic discussion, such as in this section, it helps to replace −13 by +15m/s. But for making a forecast the ambiguity does not need to be resolved. However, when interpolating in between two observed states by the weighted mean of a forward and a backward (in time) EWP forecast, the results may be ruined without solving the ambiguity (Van den Dool and Qin 1996; Van den Dool et al. 1997). Judgement is thus required. The directional ambiguity can be avoided all the way out to wavenumber 30 for both hemispheres and all seasons by using six-hourly data (which brings, however, a set of new challenges due to atmospheric tides.)

Table 3.3 complements Tables 3.1 and 3.2, and is for 500 mb height data along the equator. Perhaps we should consider ourselves lucky to get anything at all out of these calculations for the tropics because height varies very little at low latitudes and uncertainties in the analysis may overwhelm the results.

Nevertheless, one can see the long waves go westward, and at much higher phase speed, −25m/s, than observed in mid-latitudes. This agrees

Table 3.3.   The same as Table 3.1 but now along the equator. (Index m not shown)

				Zonal wave number						
>	0	1	3	5	7	9	11	units	Reference
$\langle A \rangle$ or A_{ps}	10	7	3	2	2	1	1	gpm	(3.5a)
$\langle A \rangle^2/\langle A^2 \rangle$	70	77	78	78	77	77	77	%	(3.2)
ϵ_{ps}^{+1}	ND	-19	-13	-7	4	-8	8	°λ	(3.5d)
C	ND	-25	-6	-2	1	-1	1	m/s	(3.6)
A_{ps}^{+1}/A_{ps}	98	68	60	50	43	33	20	%	(3.5a/b)

with the Rossby equation (Holton 1979) because the so-called beta effect (see Appendix 2) is largest at the equator, and also because zonal wave #1 is longer at the equator than at 50°. Because their amplitudes (first line) are so small, the phase propagation for short waves is unclear and may be close to zero at low latitudes. This too makes theoretical sense because short waves experience the zonal mean zonal wind which is strongly from the west in mid-latitudes but much weaker in a vertically integrated sense in the tropics (perhaps even weakly from the east).

Table 3.4 shows, for 50°N only, what happens when the time increment is increased from one day to two, three days, etc. For brevity only lines 4 and 5 from Table 3.1, phase speed and propagation steadiness, are shown; lines 1 and 2 are the same for all time increments anyway. For two-day increment the phase speeds are essentially the same (but steadiness decreases and ambiguity moves into longer waves).

We conclude that the quasi-linear approach is apparently valid for short time increments, ≤ two days, but the ambiguities about the direction of the phase speed penetrate towards longer and longer waves with increasing Δt. Beyond three-day separation the results fall apart. At eight-day separation (not shown) it is impossible to conclude anything. One-day increments work well—shorter increments would also be good, but extra work needs to be done on the tides.

Table 3.4.   As Table 3.1, lines 4 and 5, but for a variety of time increments.

				Zonal wave number m						
>	0	1	3	5	7	9	11	units	Reference
				$\Delta t = 1 day$						
C	ND	-3	1	5	8	10	10	m/s	(3.6)
A_{ps}^{+1}/A_{ps}	89	90	88	79	65	44	32	%	(3.5a/b)
				$\Delta t = 2 day$						
C	ND	-3	0	5	8	?	?	m/s	(3.6)
A_{ps}^{+1}/A_{ps}	70	74	70	49	23	8	7	%	(3.5a/b)
				$\Delta t = 3 day$						
C	ND	-3	0	5	?	?	?	m/s	(3.6)
A_{ps}^{+1}/A_{ps}	53	60	56	23	3	4	4	%	(3.5a/b)

Table 3.5. Tabulation of phase speed C for selected spherical harmonics of global daily 500 mb height anomalies in January as a function of zonal and total wave number n. The time increment is 24 hours. Period = 1969–1987. ND = Not Defined. Units are m/s, and the reference is Equation 3.6. The speeds shown are valid at the equator; speeds at other latitudes are obtained by multiplication by $\cos\phi$.

	Zonal wavenumber m					
>	1	3	5	7	9	11
$n=1$	−35	ND	ND	ND	ND	ND
$n=3$	−13	−9	ND	ND	ND	ND
$n=5$	−4	−3	0	ND	ND	ND
$n=7$	2	3	5	4	ND	ND
$n=9$	3	6	9	9	3	ND
$n=11$	6	9	11	13	9	3
$n=13$	7	9	11	16	14	9
$n=15$	9	9	13	16	18	13

Next we report on using spherical harmonics. For these functions, sin/cos in longitude and associated Legendre functions in latitude, we have two wavenumbers to consider. Therefore in addition to m there is also the (total) wavenumber n ($n-m$ is the number of zero crossings between the two poles). On the other hand, results are simpler in that they apply to the whole sphere at once and there is thus no need to discuss 50°N, 50°S and equator separately. Table 3.5 shows zonal (west to east) phase speeds when using spherical harmonics.

Clearly spherical harmonics obtain an even better separation in westward moving long waves and eastward moving short waves. For instance, depending on n, zonal wavenumber $m=1$ has speeds ranging from −35m/s to +7m/s. Shorter zonal waves ($m=9$), if associated with short scales in the meridional direction as well, can reach phase speeds of 18m/s in January. While meridional phase speed is not defined for either zonal or spherical harmonics, the dependence of the zonal phase speed on n (as opposed to only m) makes a major difference for wave propagation on a sphere. Theoretically (Baer 1972) the phase speed of spherical harmonics in a simple background flow depends on n only, see Appendix 2, but we find empirically a strong dependence on m as well.

3.3   Rock in the pond experiments

We are now ready for an experiment. A round disturbance is placed at 45°N on a polar stereographic map of the Northern Hemisphere, see Figure 3.1 upper left. (The values of the disturbance decrease from the maximum (150 in arbitrary units) as $\exp(-\sqrt{r/r_0})$, where r is the distance to the center in degrees and r_0 is the e-folding radius, $r_0 = 7.5$ degrees. The 20 contour

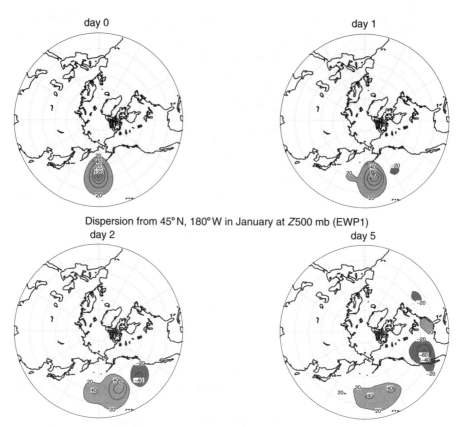

Figure 3.1 Dispersion of an isolated source, defined in Section 3.3, initially at 45°N and the date line using propagating zonal harmonics. The wave speeds are derived empirically from a multi-year 500 mb height daily data set in January. The four panels show the result after zero (upper left), one, two and five days (lower right). The geography is for orientation only. Contours every 20 gpm. Positive values light shading, negative values darker shading. Negative contours are dashed.

extends about 15 degrees from the center.) One can think of this as an isolated anomaly Z' in 500 mb height in January. The rest of the world has near-zero anomalies initially. The continental outlines are for orientation only; the experiment is zonally invariant. Contours are every 20, no zero line shown. Units (and sign) are arbitrary because the method is linear. The question is, what will happen to this initial source? If this were a passive tracer one might expect the blob to move along with the background wind. Dispersion by gravity waves would take the pressure perturbation in all directions. Here we will witness very different behavior. Decomposing the disturbance into 20 zonal waves, one can use EWP to propagate each wave at each latitude (at 2.5 degree spacing) by its own phase speed (using the complete version of Tables 3.1–3.3, all latitudes, all m) while leaving the wave amplitude unchanged, and recompose the field one day later. After one day the original disturbance has moved east, but one may notice a downstream development of opposite sign and an upstream development of like sign. At day one we have, in a sense, three rocks in the pond, each of which is repeating the process. The downstream anomaly gains amplitude by day 2 and kicks off an anomaly further downstream. One can follow the peak of the original rock moving east until day 4. A wave train $(+, -, +)$ or traveling storm track plus envelope is seen at day 3 and beyond. Remarkably, the dispersion of *stable* waves leads to downstream development and even formation of strong gradients (frontogenesis). The upstream development of the same sign causes much persistence in the area of origin. Much of the variability (75–80% of the variance) in the atmosphere can be "explained" this way (stable waves moving around). Even though in Nature one will never observe this experiment we have, by applying empiricism, found reasonable behavior and are able to demonstrate a number of physical processes. The dispersion causes the non-trivial motion of the original anomaly. For the notion "energy" think of Z'^2. This quantity is conserved in a space integrated sense, but one can see the energy travel at speeds higher than the phase speed through the wave train. This phenomenon is also called group velocity, see Holton (1979, p. 151). Comparison with numerical experiments by Simmons and Hoskins (1979) and Chang and Orlanski (1994) can be made. The shape and orientation of the eddies in the wave train is such that they would transport momentum $(u'v')$ into the jet, so remarkably a linear empirical experiment shows features of non-linearity, similar to Branstator's (1995) eddy feedback model.

Figure 3.2 is the same experiment, but now the dispersion is two dimensional by using spherical harmonics instead of (as done in Figure 3.1) zonal harmonics by latitude. From the beginning the dispersion is different in character than in Figure 3.1. While zonal dispersion can still be seen, this 2D version of EWP also shows meridional energy propagation, i.e. the anomalies travel outside the latitude band in which they were contained at $t=0$, and after a few days patterns emerge that look like veritable large-scale teleconnections. This happens even though the phase speed is always in the zonal direction. After many days the NH source even kicks off wave trains in the SH mid-latitudes (not shown). Dependence of phase speed on n thus makes a very large difference for energy propagation. (For later reference, the energy travels in a direction perpendicular to the long axis of the anomaly ellipse.) Studies of teleconnections by Rossby wave propagation on the sphere in idealized numerical models were made by Opsteegh and Van den Dool (1980) and Hoskins and Karoly (1981). Many of their results could have been obtained with the even simpler EWP approach.

In both Figures 3.1 and 3.2 one may think of the experiment in the following way. By constructive and destructive interference, a set of global functions is made to be non-zero in a local area, and zero elsewhere. As soon as the clock starts ticking, the waves move and the degree of interference

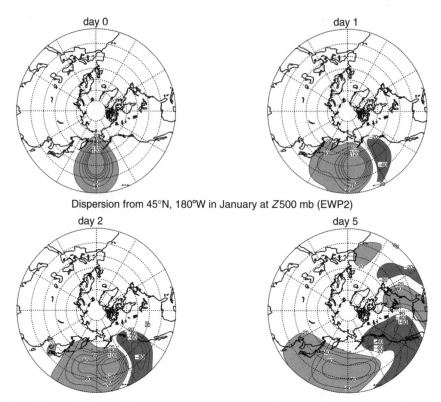

day 0   day 1

Dispersion from 45°N, 180°W in January at Z500 mb (EWP2)

day 2   day 5

Figure 3.2   As Figure 3.1. Dispersion of the same isolated source at 45°N, but now using propagating *spherical* harmonics. The wave speeds are derived from a multi-year 500 mb height daily data set in January. The four panels show the result after zero (upper left), one, two and five days (lower right). The geography is for orientation only. The contours are every 20 gpm. Positive values light shading, negative values darker shading. Negative contours are dashed. (The amplitude of the initial source is taken three times larger as in Figure 3.1 to accommodate the loss of amplitude due to dispersion over the global domain.)

changes gradually. Because of dispersion the initial blob is not just translated (to the east or west) as a single entity but shows remarkable transformation, zonally as well as meridonally.

We have essentially created a virtual laboratory experiment. The reader could place one or more sources wherever she/he wants to study propagation for a certain season and variable and then study what happens.

3.4   Skill of EWP one-day forecasts

Instead of using idealized initial states we can start from an observed anomaly field and make a 24 hour forecast which can be verified. We now discuss the skill of such EWP forecasts. The above discussion about wave energy dispersion and the basic processes of teleconnections would

Table 3.6. Anomaly correlation (x100) of 24 hr forecasts by EWP and persistence (PER) for three domains, Northern and Southern Hemisphere (NH, SH) and Tropics (TR), as a function of variable and level. Data is for 0Z, 1979–1995 in December through February. Where the gain of EWP over PER exceeds 10 points the values are underlined bold.

		Streamfunction		Velocity potential		Geop Height		Temperature	
		EWP	PER	EWP	PER	EWP	PER	EWP	PER
50 mb	NH	95	94	64	64	94	93	92	91
	TR	90	88	63	62	81	80	76	76
	SH	94	93	52	51	91	89	85	79
200 mb	NH	86	80	75	67	86	81	**74**	**63**
	TR	86	84	80	78	86	83	73	68
	SH	**86**	**75**	73	67	**85**	**72**	**74**	**53**
500 mb	NH	84	78	55	52	83	77	71	60
	TR	82	79	65	64	82	79	72	69
	SH	**84**	**70**	55	54	**83**	**69**	**74**	**53**
850 mb	NH	81	75	64	55	79	74	73	66
	TR	77	75	75	73	78	77	74	73
	SH	**80**	**70**	65	58	**78**	**67**	**71**	**53**

	Prec. water		Vert. motion 500 mb		Surf. pressure		Precipitation	
NH	**50**	**32**	16	6	76	70	15	5
TR	76	73	41	40	78	75	41	40
SH	**55**	**31**	18	6	**75**	**63**	18	6
	EWP	PER	EWP	PER	EWP	PER	EWP	PER

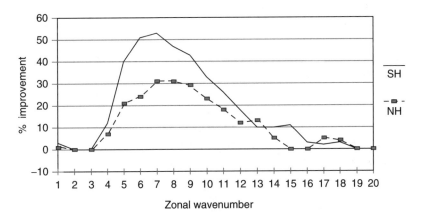

Figure 3.3 The percent improvement of rmse of one-day EWP forecasts over persistence along 50°N (dashed line) and 50°S (solid line) in DJF. For zonal waves 6 and 7 the rms error in the SH is cut in half by taking wave motion (as per EWP) into account.

be enhanced if we can show that EWP has a certain forecast capability. Because of superior performance by Numerical Weather Prediction, in 2005, EWP is not a forecast tool of practical interest. Our control to judge skill is persistence (PER), i.e. persist yesterday's anomalies for 24 hours as a "lazy man's forecast". Both EWP and PER are verified over a very large number of cases using the anomaly correlation defined in Chapter 2. Results are given in Table 3.6

For all variables and levels EWP is better than PER. This is true even for every single forecast. Indeed EWP is a very safe and conservative forecast (wave dispersion is always in effect). The gains over PER are largest in the troposphere, the mid-latitudes, the southern hemisphere and for temperature. The largest gains are about 20 points. In the tropics, in the stratosphere, and for velocity potential, EWP has very little skill over PER at day 1. At the bottom of Table 3.6 some assorted variables are listed. The surface pressure behaves consistent with heights aloft. In fact, phase speeds are nearly constant from sea level to 50 mb (not shown) as systems appear to travel with strong vertical coherence. Vertical motion and rainfall are nearly impossible to forecast, but even here some propagation can be surmised. Precipitable water in the atmosphere is not easy to forecast either, but EWP does have a large gain over PER.

We conclude the EWP describes realistic processes because it results in forecasts with a substantial gain in skill over persistence. If the reader feels that EWP is like a barotropic model, please note that EWP works equally well for many variables at many levels, not just $Z500$ in mid-latitudes.

Figure 3.3 shows the reduction in root-mean-square error (rmse) of EWP relative to PER as a function of wavenumber for the NH and SH along 50°. It is pretty obvious that the gains are due to mobile waves $m = 4$–13. Without taking wave motion into account (PER) the error is large, while EWP accomodates the motion (if only in an averaged sense), cutting the rmse by up to 50%. In the long waves EWP does not beat PER, even though the phase speed is non-zero. This is because it takes a high speed for a long wave to travel an appreciable distance relative to its own wavelength (which is what is needed to beat PER). Such speeds are not observed. The $+10$m/s for short waves is worth a lot more in terms of forecast skill than the -25m/s for the longest waves. Figure 3.3 is for EWP using zonal waves. Use of spherical harmonics lowers forecast skill everywhere! Apparently there is some merit in localizing the phase speed estimates.

3.5   Discussion of EWP

3.5.1   Eulerian and Lagrangian persistence

Table 3.7 highlights that EWP has virtually the same score under all circumstances. For winter or summer, southern or northern hemisphere, the EWP score is always around 0.82.

The spatial variation in EWP verification scores is also very small, see Qin and Van den Dool (1996), their Figure 3. It is quite unusual for a forecast

Table 3.7. Scores (anomaly correlation) of one-day forecasts by EWP and PER in the two high seasons in the two extratropical hemispheres for 500 mb height.

	EWP	PER
DJF-NH	0.81	0.73
DJF-SH	0.83	0.64
JJA-NH	0.84	0.77
JJA-SH	0.83	0.67

technique to have the same skill everywhere all the time, a primary example being PER which has much higher scores in the NH. While PER is Eulerian persistence (skill of PER is low when systems move very fast), EWP may be described as a first-order attempt to measure Lagrangian persistence. After all, the similarity to the Rossby equation, too good to be a coincidence, suggests we have empirically solved the problem of following parcels that conserve absolute vorticity. This principle is apparently equally valid under all circumstances.

3.5.2   Reversing time and targeted observations

One may wonder how one would ever get an initial blob at 45°N, as in Figure 3.1. The answer is obtained by flipping over a transparency of Figure 3.1 and thinking of $+n$ days as $-n$ days. A constellation of positive and negative upstream perturbations at -5 days all collapse into a single positive perturbation at $t=0$, using EWP propagation speeds in reverse. A modern interpretation is related to the topic of targeted observations. Suppose one wants to make the (NWP) forecast in the area of the blob better; where should additional observations n days ahead of time be taken? If uncertainty in the upstream sensitive areas can be reduced by additional observations, the amplitude of the uncertainty blob (to be placed at the location of interest) can be greatly reduced n days later. To paraphrase Joe Tribbia at the AMS 2005 meeting: "Group velocity gives a wave dynamics perspective to adaptive observations strategies". Subject to assumptions of reversibility, linearity, etc., EWP can be used for a most simple explanation of the idea of targeted observations. For complex models, writing a linearized version that can be integrated back and forth in time is difficult. EWP is linear and goes back and forth in time as constructed.

3.5.3   Application of EWP

It would be a stretch to believe that EWP has wide practical application in forecasting. Its use is mainly for teaching and demonstration purposes. Nevertheless there are a few applications with practical meaning based on EWP, which we list here.

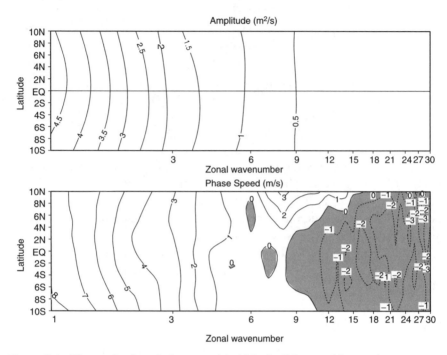

Figure 3.4 The amplitude and phase speed (m/s) in April for zonal harmonic waves, as a function of zonal wavenumber and latitude, in 200 mb velocity potential as determined from 12 hourly data for the period 1979–2003. The zonal wavenumber (1–30) is on a log scale. Unit for the amplitude is $10^6 m^2/s$. Negative values are shaded. Negative contours are dashed.

(a) NWP models have been weak at some specific features, such as the Madden and Julian Oscillation (MJO; Madden and Julian 1971), a global phenomenon in the tropics traveling eastward along the equator (Waliser et al. 2006). Real-time forecasts of the MJO by several methods including EWP can be found on the web. In the case of MJO, EWP was applied to 200 mb velocity potential fields; see link at http://ww.cpc.ncep.noaa.gov/products/precip/CWlink/MJO/mjo.html, and the article under readme. In spite of the strong westward speed reported for the height field (or streamfunction) in the tropics, see Table 3.3, the longest waves in velocity potential anomalies travel eastward with a speed that varies semi-annually between 5.5m/s (February and September) and 11m/s (May and November). These eastward speeds may indicate the dominance of Kelvin waves. Figure 3.4 gives the near-equatorial phase speeds of 200 mb velocity potential anomalies for waves 1–30 in April. Long waves move eastwards, but short waves move westwards. The longest waves have the largest amplitude, and the amplitude is highest near the equator. (b) Interpolation of weather maps which are provided infrequently. EWP is much more accurate at interpolation than linear interpolation, see Van den Dool and Qin (1996). This EWP application will exist as long as model forecast and analysis results are provided infrequently (∼ every 6 hours). For potent small-scale moving weather systems EWP is a good interpolation method even if weather maps were available every hour. EWP can likewise be applied to provide time interpolated boundary conditions for a limited area model, given global forecasts every 6 hours.

(c) In many geophysical disciplines there is a need to know atmospheric tides at high temporal resolution. Interpolation of very fast atmospheric tides (one revolution per day) from 6 hourly data to once hourly was addressed in Van den Dool et al. (1997) and the results are applied in several geodetic and oceanographic research areas. Salvaging data originally sampled at the Nyquist frequency (the semidiurnal tide dominates!) is a peculiar challenge that can be addressed using EWP.

(d) The guess field for data assimilation: in data assimilation an NWP model is used to make the guess field. This is a drawback in case the model keeps introducing certain systematic errors. EWP could be used to advance the previous analysis for 3 or 6 hours, i.e. EWP could make the guess field.

3.5.4 Historical note

EWP could have played a role in the history of objective forecast methods in the 1950s and 1960s, and be competitive with numerical models at least until the first hemispheric baroclinic models emerged around the mid-1960s. But it did not. Attempts to apply the theoretical Rossby equation wave by wave are known only by anecdote but were given up on soon. This may have been because the Rossby equation has unrealistic speeds for long waves, a problem EWP (which was unknown, although Eliasen and Machenhauer (1965) comes close) does not have. (EWP does not require the so-called Cressman correction.) EWP would have been difficult to apply because hemispheric data without gaps were not available routinely until the limited area models had been replaced by hemispheric models. EWP is hindered greatly on a limited area (like the ocean), because the treatment of the phase of sine/cosine waves at the boundaries is difficult. The empirical evidence that the Rossby equation has validity was usually phrased in terms of group speed, for instance by pointing at the propagation in physical space of a trough in a so-called Hovmoller diagram (Hovmoller 1949; Holton 1979, p. 151). Madden (1978), using stringent criteria, listed periods in the historical record when Rossby waves were clearly present (less than 20% of the time). Roads and Barnett (1984) made a "dynamically oriented statistical forecast model", but found little forecast skill beyond damping and persistence; in particular they mention that a regression involving projections on sine and cosines did not pick any mixed predictors indicating an absence of systematic propagation. Phase shifting was conceived of in a purely intuitive fashion, in studies like Cai and Van den Dool (1991;1992). In Cai and Van den Dool (1994), where composites were made in a framework following a large-scale wave, it became obvious from their Figures 2–5 that phase propagation *à la* Rossby is evident in the data. The EWP presented here is the simplest form of time averaging phase shifted data.

3.5.5   Weak points of EWP

We hope the reader feels encouraged to interrogate a data set to come up with empirically based methods that can be used to explain difficult concepts, and to some degree to make forecasts. Certainly, EWP is not perfect. In listing weak points we hope some readers will be inspired to try to outdo something as deceptively simple as EWP.

(1) EWP derives propagation properties without regard for longitude, i.e. for Figure 3.1 it makes no difference where the source is situated relative to longitude or the standing waves (implicitly the land–ocean distribution).
(2) Propagation speeds are independent of time (except the annual cycle). In view of the Rossby equation one might want to study time variation on the interannual time-scale. In years with stronger jets, waves should move faster.
(3) The meridional scale was neglected in EWP1 (zonal harmonics), and when it was included (EWP2, spherical harmonics) the forecast skill decreased. This appears to be because spherical harmonics give estimates as a compromise between two hemispheres (in different seasons), tropics and mid-latitude. An approach in a restricted latitude bands, like 20°N to the pole, with sine waves in the north–south direction (semi-Fourier) would improve upon EWP1.

At a more fundamental level and beyond repair is the drawback that EWP does not deal with energy lost to or gained from the basic state. As an arching route is followed by a group of disturbances new energy added to the disturbance may be as large as the energy from the initial source. In Chapter 7 we will see that an empirical method called constructed analogue makes forecasts *à la* EWP, but some interaction with the basic state can be seen.

Appendix 1: EWP formal derivation

One can derive EWP formally by asking which constant propagation angle ϵ and damping factor ρ should be applied to a wave at $t=0$ so as to yield, on average over many cases, the best forecast of that wave at a future time. Here, best forecast means lowest rms error on average. The observed (phase shifted) waves at $t=0$ and $t=1$ being $A \cos x$ and $c \cos x + d \sin x$, respectively, and the forecast for $t=1$ based on observations at $t=0$ being $\rho A \cos(x - \epsilon)$, one thus needs to minimize

$$U = \sum (\rho A \cos\epsilon - c)^2 + (\rho A \sin\epsilon - d)^2. \tag{3.A1}$$

In (3.A1) summation is over all times, A, c, d are a function of time (with non-zero means), and ϵ and ρ are constant in time. Upon differentiation of U wrt ρ and ϵ one obtains:

$$\sum \rho A^2 - c\, A \cos\epsilon - dA \sin\epsilon = 0$$

and

$$\sum c\, A \sin\epsilon - d\, A \cos\epsilon = 0.$$

The solution (to be evaluated from a data set) is:

$$\epsilon = \mathrm{atan}\{\langle dA\rangle/\langle cA\rangle\} \tag{3.A2}$$

where $\langle\ \rangle$ are time means. Given ϵ, the damping ρ can be calculated as

$$\rho = [\langle cA\rangle \cos\epsilon + \langle dA\rangle \sin\epsilon]/\langle A^2\rangle. \tag{3.A3}$$

Equation (3.A2) can also be written as

$$\epsilon = \mathrm{atan}[\{\langle d\rangle\langle A\rangle + \langle d'A'\rangle\}/\{\langle c\rangle\langle A\rangle + \langle c'A'\rangle\}]. \tag{3.A4}$$

If we neglect the covariance between A and both c and d, the requirement for minimal U is $\epsilon = \mathrm{atan}\{\langle d\rangle/\langle c\rangle\}$, which is EWP as proposed intuitively, see Eq (3.5d). Even if the transients terms like $\langle d'A'\rangle$ are not small the resulting ϵ may not necessarily differ much from $\mathrm{atan}(\langle d\rangle/\langle c\rangle)$. An evaluation on the data used in Tables 3.1–3.3 shows extremely minor influence from the covariances on the resulting ϵ. The impact would be larger if, for instance, above average A is associated with faster than average propagation; apparently this does not happen.

Note that damping the wave, $\rho < 1$, for the purpose of lowering rms error does not change the requirements for optimal propagation. The damping ρ, given by (3.A3), goes to zero for large time increments (say 10 days).

Two final comments:

1. One can derive EWP formally with all the result shown above without phase shifting! The expressions are a little longer but otherwise the same. So while phase shifting is helpful, it is not necessary.
2. The rms minimization for phase shift and amplitude damping appear unrelated, i.e. one can simplify the exercise of finding ϵ without mention of ρ.

Appendix 2: The Rossby equation

The simplest expression for the phase speed of Rossby waves is $C = U - \beta/K^2$, where C is the phase speed, U is the background wind speed, β is the meridional derivative of the Coriolis parameter, and K is the wavenumber (if only the zonal wavenumber is considered K relates to m as follows: $K = 2\pi m/L$, where L is the circumference of the Earth). This was derived by Rossby and collaborators in 1939. A more complete expression is $C = U - (\beta - \partial^2 U/\partial y^2 + F^2 U)/(K^2 + F^2)$, where K is the three-dimensional wavenumber, F is the inverse of the Rossby radius of deformation, and $\beta^* = \beta - \partial^2 U/\partial y^2$ is the apparent β-effect. Ignoring the vertical wavenumber, and setting the meridional wavenumber l equal to $n\pi/a\cos\varphi$ (and taking $n=3$, independent of m) the wave speeds at 50°S and 50°N in Tables 3.1 and 3.2 are actually to within 1.5 m/s of theory, i.e. in quantitative agreement with the Rossby equation. Likewise we found the seasonal cycle in phase speed along 50°N/S to be well explained by the seasonal cycle in U, and β^*.

The energy travels with the group speed (c_{gx}, c_{gy}) which can be derived from the above by differentiation wrt wavenumber. We find

$$c_{gx} = U + (\beta^* + F^2 U)(k^2 - l^2)/(K^2 + F^2)^2 \text{ and } c_{gy} = 2\beta^* (kl)/(K^2 + F^2)^2,$$

see Pedlosky (1979, p. 114, Chapter 3). Clearly group speed is higher in the zonal than in the meridional direction; see also the comments in Chapter 4 about the shape of eddies.

As for an expression on the sphere we obtained

$$C = U_{eq} - \frac{2(\Omega a + U_{eq}) + a^2 F^2 U_{eq}}{(n(n+1) + a^2 F^2)}$$

where Ω is the rate of rotation of the Earth, and a is the radius of the Earth. This expression holds only for a simple background flow $U(\phi) = U_{eq}\cos\phi$, also called super-rotation. The phase speed given is valid at the equator. Interestingly, phase speed depends theoretically on n only. However, in Table 3.5 we established empirically that phase speed depends nearly as much on m.

4 Teleconnections

Mankind has long been intrigued by the possibility that weather in one location is related to weather somewhere else, especially somewhere very far away. The fascination may be mostly related to possible predictions that could be based on such relationships. The severe weather that harmed the British Army in the Crimea in November 1854 (Lindgrén and Neumann 1980) was due to a weather system moving across Europe, suggesting it could have been anticipated from observations upstream. It took analyses of many surface weathermaps, an activity starting around 1850, to see how weather systems have certain horizontal dimensions, thousands of kilometers in fact, and move around in semisystematic ways. It thus followed that, in a transient sense, the weather at two places can be related, and in a time-lagged sense that weather observed at one (or more) places serves as a predictor for weather at other locations. The other reason for fascination with teleconnection might be called "system analysis". The idea that given an impulse at some location ("input") a reaction can be expected thousands of miles away (the "output") through a chain of events, is intriguing and should tell us about the workings of the system. It is akin to an engineer testing electronic equipment. Unfortunately, Nature is not a laboratory experiment where we can organize these impulses. Only by systematically observing what Nature presents us with, may we dare to search for teleconnections in some aggregate way.

The word teleconnection suggests a connection at long distance, but a stricter definition requires some thought and pruning down of endless possibilities. We need to make choices about (a) simultaneous vs time-lagged teleconnections, (b) correlations vs other measures of "connection", (c) transient vs standing teleconnections, (d) teleconnections in filtered data (e.g. seasonal means) vs unfiltered instantaneous (e.g. daily) data, and (e) one or more variables. On (a), (b) and (e) our choice in this chapter is simultaneous,[1] use of linear correlation (except in section 4.3 where other measures of teleconnection are discussed), and a single variable respectively. On possibilities (c) and (d) we keep our options open.

---

[1] In doing so we realize that teleconnections defined as simultaneous have no predictive value as such. We come back to the prediction issue at the end of this chapter.

4.1   Working definition

A teleconnection is a simultaneous significant[2] temporal correlation in a chosen variable between two locations that are far apart, where "far" means beyond the monopole of positive correlations that is expected to surround each grid point or observational site. "Beyond the local positive monopole" implies we should first look for significant negative correlation, keeping in mind there may be significant positive correlation at even greater distance. These teleconnections should exist in the original "raw" data and in that sense be "real". By far the two most famous teleconnections in the extra-tropical NH are the North Atlantic Oscillation (NAO) and the Pacific North-American Pattern (PNA). The most important teleconnection with predictive implications, is probably the global ENSO teleconnection.

4.2   Two most famous examples in NH

A few examples should focus the discussion. Figure 4.1 shows two patterns that most experts will identify as the NAO and PNA. They were calculated from seasonal mean (JFM) 500 mb height for the period 1948–2005, a total of 58 realizations of seasonal mean flow for the area north of 20°N. The two maps are of the "one-point teleconnection" variety, terminology due to Wallace and Gutzler (1981), i.e. for the NAO we have chosen the base point at 65°N, 50°W, and for the PNA at 45°N, 160°W. How we came to choose these base points will be discussed later. The maps show contours of the correlation between time series at the base point and all other points. In view of (2.14) the correlation ρ_{ij} (shorthand for $\rho(s_i, s_j)$ is given by

$$\rho_{ij} = q_{ij}/\sqrt{q_{ii} \cdot q_{jj}}, \tag{4.1}$$

where i is the base point and j are all other points, $1 \le j \le n_s$. The NAO pattern shows a positive correlation around its base point, as expected around any base point, but more interestingly, a very large area of negative correlation to the south stretching from North America to deep into Europe along 35–45°N (sloping northward as one goes east). This means that when 500 mb height is higher than usual near southern Greenland it tends to be lower than normal along 40°N and vice versa. This "see-saw" in 500 mb height, by geostrophic approximation, modifies the strength of the westerlies (or polar jet) in between the two main centers of the NAO across the Atlantic. Further to the south, 20–30°N, heights are positively correlated with the Greenland base point. In the phase where the polar jet is strengthened, the subtropical jet is weakened, and vice versa. In a nutshell,

---

[2] Here "significant" means both statistically significant and of practical importance.

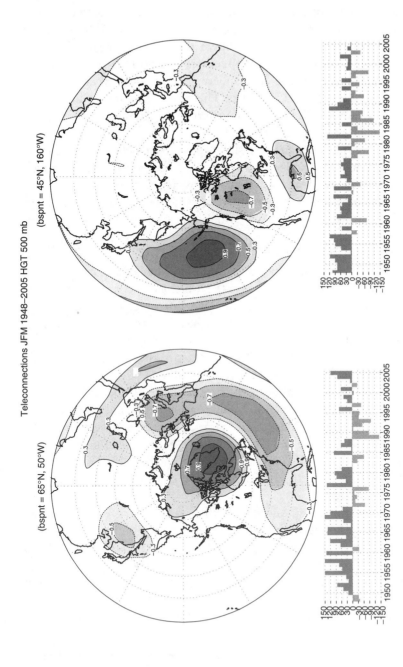

Figure 4.1 (see Plate 1) Display of teleconnection for seasonal (JFM) mean 500 mb height. Shown are the correlation between the base point (noted above the map) and all other grid points (maps) and the time series of 500 mb height anomaly (geopotential meters) at the base points. Contours every 0.2, starting contours ± 0.3. In both maps and time series the color red (blue) is used for positive (negative) values. Data source: NCEP Global Reanalysis. Period 1948–2005. Domain 20°N–90°N. On the left a pattern referred to as the North Atlantic Oscillation (NAO). On the right the Pacific North American pattern (PNA).

this is the most famous teleconnection in the NH, discovered and named by Walker (1924),[3] who worked with mean sea-level pressure data. The NAO is a standing oscillation; there is no implied motion of the pattern, just a change in polarity described by the sign of the time series of 500 mb height anomalies at 65°N, 50°W, which is shown directly underneath the map. There are a few scattered weaker centers of positive and negative correlation elsewhere over the hemisphere, but they are weak and only the one near coastal east Asia is robust. The NAO has a strong link to the alternation of westerly and blocked flow across the Atlantic and is present from the surface up into the stratosphere.

The map in the right in Figure 4.1 shows positive correlations close to its chosen base point at 45°N, 160°W, as expected, but now negative correlations both "upstream" near Hawaii and "downstream" over west-central Canada. Furthermore, there is a positive correlation over South-East North America. In contrast to the NAO, which has two (maybe three) main centers, the PNA has four main centers. The PNA centers are organized along an arching pattern, looking somewhat like EWP dispersion in two dimensions (see Chapter 3, Figure 3.2), and therefore is suggestive of wave energy traveling from the HI center, via the North Pacific, and from central Canada to South-East North America. (We infer the direction because the group speed of Rossby waves always has an eastward component.) Extrapolating further upstream the wave energy appears to come from the deep tropics near the date line.[4] In contrast to the Atlantic the Pacific has only a subtropical jetstream in the mean, and the PNA does modify this jet, but mainly west of 150°W. As is the case with the NAO, the PNA has a clear overlap with the phenomenon of alternating periods of blocked flow, particularly in the Gulf of Alaska, and periods of stronger westerlies. The PNA was named by Wallace and Gutzler (1981), but can be found without a problem in the atlases of O'Connor (1969) and Namias (1981). Just as the NAO, the PNA is a standing oscillation which changes polarity, but does not appear to propagate.

The time series, the height anomaly at the base points, are often studied for long-term trends. Indeed, the PNA time series suggests that the polarity opposite to what is shown in Figure 4.1 was uncommon before 1976 (Douglas et al. 1982; Trenberth 1990). The trend towards negative in the NAO time series from the 1950s to 1990s was noted also by Hurrell (1995) and linked to higher temperatures over the Eurasian continent in winter and even the global mean temperature. However, this trend has since

---

[3] Perhaps Walker would not agree with this assessment because the alternation of strong westerlies and blocked flow at the time-scales of weeks was well known in Europe in the nineteenth century, see Rogers and van Loon (1978).

[4] This suggestion about tropical forcing explains the explosion in popularity the PNA received in the mid-1980s when a global observing capability in real time was in place for the first time during a strong El Nino event (1982–83).

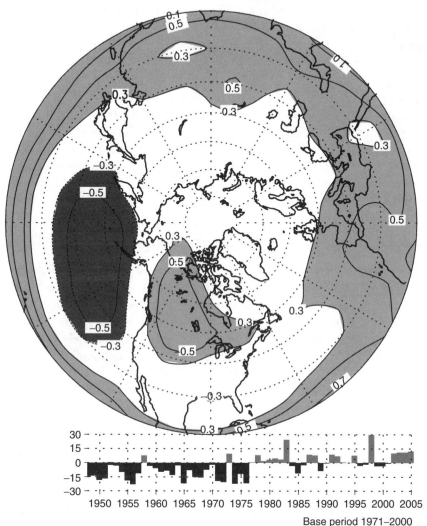

Figure 4.2 Same as Figure 4.1, but now the base point is at 2.5°S and 170°E, i.e. outside the display area. Positive values light shading, negative values darker shading. Negative contours are dashed.

faltered. The main variation in Figure 4.1 is inter-annual, not a long-term trend.

It should be noted that the PNA and NAO operate in nearly distinct spatial domains. In other words, in view of Equation(2.1) these two patterns are nearly orthogonal. This happens in a natural way, not by mathematical design, because calculating teleconnections this way has no orthogonality requirements built in. The only overlap is over eastern

North America and the adjacent Atlantic, where the PNA and NAO may be in competition.

Figure 4.2 shows a rendition of the teleconnection that currently is most famous of all. We place the base point at 2.5°S, 170°E (i.e. outside the domain displayed) and calculate the correlation of 500 mb height in the deep tropics near the date line with grid points over the NH (20°N–pole). At this point we do not invoke SST or tropical forcing explicitly. We just note that when heights are higher than average near the date line, heights tend to be higher than normal everywhere in the tropics (not shown), and into the subtropics and lower mid-latitudes, i.e. positive correlation over an area covering nearly half the planet. In the north of the general Pacific North American area we find negative correlation near the Aleutian Islands, and positive over North-West Canada. Study of the correlation of the tropics with the extra-tropics in the PNA area was pioneered by Horel and Wallace (1981), at a time when much less data were available. The positive excursions in the time series in Figure 4.2 mark the years of all the famous El Ninos (1958, 1973, 1983, 1998). However, the time series also has a dominant upward trend, or perhaps a discontinuity near 1977, which serves as a reminder that researchers have to decide whether this is real (faithful to nature), or caused by inhomogeneities in observations used in the NCEP/ NCAR Reanalysis. Another point of discussion is whether Figure 4.2 shows the PNA in mid-latitudes. After more than a decade of loosely calling the mid-latitude portion of Figure 4.2 the PNA, there is enough of a shift in space to consider the pattern associated with ENSO events in the tropics to be different from the PNA (Livezey and Mo 1986; Straus and Shukla 2002). The reader can study this by comparing Figures 4.1 and 4.2. The ENSO teleconnection modifies the subtropical jet in the Pacific, but farther east than the PNA does. Other estimates of the ENSO teleconnection will be presented in Chapters 5 and 8.

4.3 The measure of teleconnection

Teleconnections have been studied primarily with linear correlation, as in Equations (2.14) and (4.1). We did the same in the above, but there are different techniques that have been used, are implicit in EOFs, and should be considered for better understanding. Instead of correlation, one can use its close companion, the regression coefficient. Plotting teleconnections using regression coefficients has been rare because the remote centers of opposite sign generally look less prominent that way and thus less interesting, especially in the Pacific. (Regression coefficients are useful in making orthogonal functions called EOT, see below.) Prominent among the alternatives are "composites" based on satisfying a particular criterion. For instance, a composite mean for all cases when the anomaly at the North Atlantic base point of the NAO is greater (smaller) than $\pm\gamma$ times the local standard deviation. The threshold factor γ, say 0.5, can be changed, but one needs to retain enough samples. The two composites provide a test as to the validity of the assumption of linearity which underlies correlation (or covariance). Table 4.1 gives an impression of these different options for our choice of two grid points. We here take the time series of JFM mean 500 mb height at 65°N, 50°W ("Greenland"), as shown in Figure 4.1, and at 47.5°N, 5°E ("Europe") for 1948–2005. These points closely represent two centers of the NAO (for the southern center there are many other good choices). Their correlation is −0.67, indicative of the see-saw. Given 58 years (58 samples, assuming each year is independent) there is less than 5% chance of a correlation beyond ±0.27 if in truth the correlation is zero. This is no more than a background comment because all assumptions are violated because we have searched for and selected these points for having a high negative correlation.

The correlation between two points ρ_{ij} is symmetric (by design), and suggests a −0.67 (in units of standard deviations) 'forecast' at one point when a one standard deviation anomaly is observed at the other. In contrast, the regression coefficient is not symmetric. Because the standard deviation (72.5 vs 53.6 geopotential meter) is much higher near Greenland than over Europe, the regression coefficient $a_{ij} = (\rho_{ij}\sqrt{q_{jj}}/\sqrt{q_{ii}})$ is lower than the correlation coefficient (−0.50 vs −0.67) when choosing Greenland as base point. This is why fewer authors would show it that way, because it makes the NAO look like a northern monopole only. On the other hand, a regression from Europe to Greenland has considerably higher regression coefficient, which is, cosmetically, better for a demonstration of tele-connection. Nevertheless correlation and regression are based on the same information.

More importantly, there is the possible asymmetry not captured in either correlation or regression. The composites for a threshold of half a standard deviation, based on 16–22 cases, indicate that the linear correlation is supported, to first order, in equal parts by data points with negative and positive sign, see Table 4.1. For example, a value above the threshold in Greenland is associated with, on average for 17 cases, −0.73 local standard

Table 4.1. Listing of different ways of characterizing the teleconnection between 65°N, 50°W and 47.5° N, 5°E (Europe).

	Greenland to Europe	Europe to Greenland	units
Correlation:	−0.67	−0.67	non-dimensional
Regression coefficient:	−0.50	−0.90	non-dimensional
Composite: $> 0.5^*$sd	−0.73 (17)	−0.87 (17)	standard deviation
Composite: $< -0.5^*$sd	+0.79 (20)	+0.65 (22)	standard deviation

deviations in Europe, while a value under the negative threshold in Greenland is associated with +0.79 over Europe. That is very close to symmetry or linearity. In this case the correlation is a satisfactory tool, and more accurate than a composite because a bigger sample is used for the correlation. However, the base point composites using a threshold in Europe indicate that a response over Greenland is stronger for positive than negative anomalies over Europe. Given the small sample size this asymmetry may not be significant. For any formal test one needs to take into account the different sample sizes among the composites (this changes when varying γ), and the possibility of skewed distributions for the height field (White 1980).

4.4   Finding teleconnections systematically: Empirical Orthogonal Teleconnections (EOT)

Using Equation (4.1) Namias (1981) evaluated the correlation between any base point and all other points, i.e. for each value of i, there is a map of ρ_{ij}, $1 \leq j \leq n_s$. Since i can be varied from 1 to n_s as well, one has a full atlas of n_s one-point teleconnection maps. Namias (1981) provides such an atlas for the four main seasons for 700 mb height, 200 pages in all. His work was an update and extension of an atlas by O'Connor (1969). O'Connor's atlas in fact consisted of composites of the NH 700 mb[5] height field given that at one particular point the anomaly is in excess of some threshold, i.e. asymmetry between positive and negative anomalies was surveyed.[6] Both O'Connor and Namias had a practical application is mind. In long-range forecasting one would encounter the situation of being relatively certain about the forecast at one or at most a few points in the NH, and the task was to sketch the rest of the field (by hand in those days) using the teleconnection atlas. To this day the CPC is following this process in making the 6–10 day and week 2 forecasts and has an updated electronic version of O'Connor's atlas to do this work; the most recent reference is Wagner and Maisel (1989).

In spite of these early efforts with a practical application, there is at first sight limited science in calculating ρ_{ij} endlessly (more output than input). The effort to systematize these calculations with the purpose of finding just

---

[5] The reader may wonder why 700 mb was chosen originally in the 1940s. The interest in some mid-tropospheric level where the barotropic model could be applied with the most success led to a difference of opinion among Europeans (500 mb) and North Americans (700 mb). Once committed to the analysis at these levels the tradition continued until Reanalysis allowed researchers to chose vitually anything they want.

[6] Actually, O'Connor (1969) had an added condition, namely that the grid point had to be a center of a larger scale anomaly. This reduces the amount of data one can work with, and over time this extra condition was dropped.

the main (very few) teleconnections started with Wallace and Gutzler (1981). They searched for those base points s_i, that have the strongest negative correlation with some remote point s_j (and usually with an area around s_j). They summarized their findings on a teleconnectivity map indicating areas that relate (with negative or positive correlation) to other remote areas. The two patterns in Figure 4.1 are a summary as well in that we picked the two best-situated base points s_1 and s_2. Keep in mind that one also gets the PNA by taking a base point near HI, or Canada, but these are the redundant doubles. A third and fourth pattern can be displayed, but they explain much less variance (a topic not well developed for one-point teleconnections because orthogonality is not enforced) and are far more sensitive to adding or subtracting a year in the data set. The PNA and NAO are robust and not too sensitive to adding or subtracting a few years, and can be found by every reasonable technique.

A weak point of teleconnections á la Wallace and Gutzler (1981) is that one cannot easily (by projection) represent the original data in terms of a linear combination of NAO, PNA, etc. Both patterns are derived straight from the original data, as opposed to deriving the second pattern after the first was removed from the data. The latter can be done by orthogonalizing the base point teleconnection approach. Given the first point of choice (for whatever reason) s_1, one can reduce the anomaly data by

$$f^{\text{reduced}}(s,t) = f(s,t) - a(s_1,s)\, f(s_1,t),$$

where $a(s_1, s)$ is the regression coefficient between s_1 and any other point s. Then the next task is to find a second point s_2 (by whatever criterion) in the reduced data, and so on for the third point, after reducing the data a second time. It is easy to see that the temporal correlation between $f(s_1, t)$ and $f^{\text{reduced}}(s_j, t)$ is zero for all j. Because of this orthogonality (in time) this procedure allows functional representation as per (2.7a) as follows:

$$f(s, t) = \langle f(s, t) \rangle + \sum_{m=1}^{M} a(s_m,s)\, f(s_m,t) \quad 1 \le s \le n_{\text{s}},\ 1 \le t \le n_{\text{t}} \quad (4.2)$$

where $a(s_m, s)$ and $f(s_m, t)$ are derived from $m-1$ times reduced data. In addition to functional representation one can now also define the notion explained variance by each teleconnection pattern. In fact explained variance gives the most rational basis for choosing s_1, s_2, etc. in a certain order. One wants to maximize $\text{EV}(i)$, i.e. find that s_i for which

$$\text{EV}(i) = \sum_{j=1}^{n_{\text{s}}} \rho_{ij}^2 \times q_{jj} \quad (4.3)$$

is the highest.

Figure 4.3 shows a map of $\text{EV}(i)$, $i = 1, \ldots, n_{\text{s}}$, for the JFM 500 mb data. On the upper left one can see several areas where time series at points

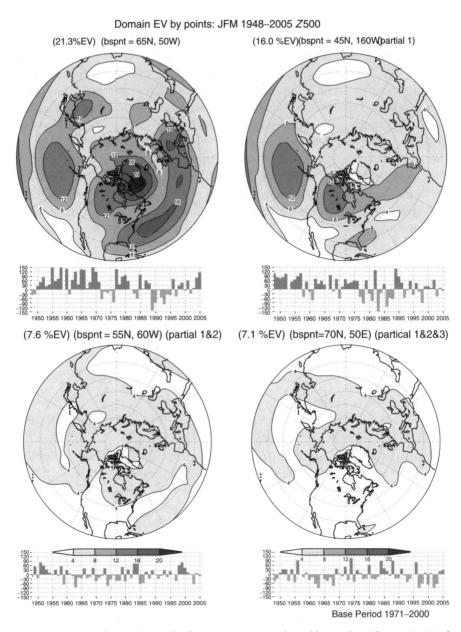

Figure 4.3 (see Plate 2) EV(*i*), the domain variance explained by single grid points in % of the total variance of seasonal (JFM) mean Z500 over 1948–2005, using Equation (4.3). In the upper left for raw data, in the upper right after removal of the first EOT mode, lower left after removal of the first two modes, etc. Contours every 4%. Values in excess of 4% lightly shaded, in excess of 12% dark shading. The time series shown are the residual height anomaly at the grid point that explains the most of the remaining domain integrated variance.

explain more than 16% of the variance at all other points combined. The points in these areas are those associated with the NAO and PNA. The highest EV(*i*) is 21.3% at 65°N, 50°W. Picking this point leads to a description of the NAO. After reducing the data set once the new EV(i) map on the upper right emerges, and the PNA is the obvious next choice. After removing the PNA a rather bland field of EV(i) at 8% or less remains (see lower left), and the choice of the next point is a moot point and depends sensitively on adding or subtracting a year from the data set, or changing the domain somewhat. Nothing stands out beyond NAO and PNA.

If one follows the procedure described above to pick s_1, s_2, etc. one obtains functions named Empirical Orthogonal Teleconnections, see Van den Dool et al. (2000), or at least one version of them. Figure 4.4 is like Figure 4.1 but presented as EOTs. The two first base points in both Figures 4.1 and 4.4 were chosen by maximizing explained variance as per Equation (4.3). While the first two patterns in Figures 4.1 and 4.4 look very similar, there are these differences in methodology and display:

(1) Figure 4.1 has correlation; Figure 4.4 has regression $a(s_m, s)$.
(2) Figure 4.1 is derived from full original anomaly data, while in Figure 4.4 the m'th pattern is derived from $m-1$ times reduced data.
(3) Teleconnections are chosen for the existence of remote negative correlation, while EOTs include a premium for explaining variance nearby, i.e. nearby positive correlation adds to EV(i).
(4) Figure 4.4 is consistent with Equation 4.3 and the notion "explained variance" now has a meaning.

In spite of these differences, EOT still resembles the well-known one-point teleconnection patterns, at least for the first few modes, but has the advantages of functional representation and explained variance. EOT are much like EOFs (next chapter), and are almost indistinguishable from the most common type of rotated EOF (Smith et al. 2003).

4.5 Discussion

There is a large body of literature since the early 1980s that attempted to study teleconnections via EOFs, but EOFs nearly always had to be rotated (Horel 1981; Barnston and Livezey 1987) so they would better resemble the Wallace and Gutzler one-point correlations. Barnston and Livezey (1987) gave an exhaustive classification of teleconnections, using rotated EOF, well beyond just NAO and PNA, for all 12 months of the year. EOTs are much simpler to calculate than rotated EOF, with no truncation and rotation recipe required.

Much has been made in the literature over the past 25 years about the shape and orientation of "eddies". The summary is that low-frequency eddies are more often identified as zonally elongated, while the high-frequency

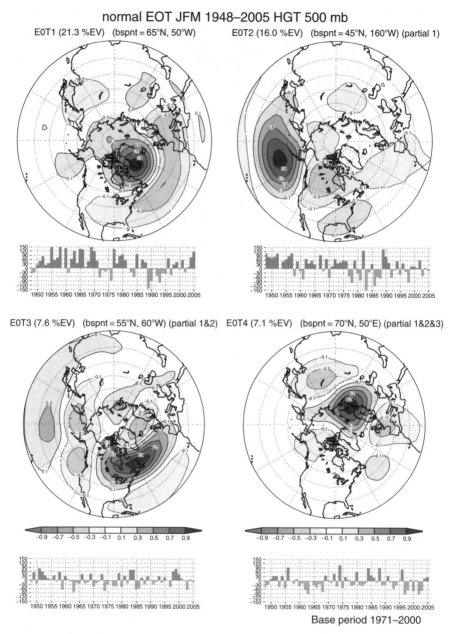

Figure 4.4 (see Plate 3) Display of four leading EOT for seasonal (JFM) mean 500 mb height. Shown are the regression coefficient between the height at the base point and the height at all other grid points (maps) and the time series of the residual 500 mb height anomaly (geopotential meters) at the base points. In the upper left for raw data, in the upper right after removal of the first EOT mode, lower left after removal of the first two modes, etc. Contours every 0.2, starting contours $\pm$ 0.1. Data source: NCEP Global Reanalysis. Period 1948–2005. Domain 20°N–90°N. (A light post-processing was applied, See Appendix I of Chapter 5)

eddies are more often meridionally elongated. Eddies in different frequency bands are typically obtained by applying a digital filter to the observations. In this chapter the examples always used seasonal mean data, thus emphasizing zonally elongated eddies suggestive of meridional energy transport. At the very least the PNA looks very much like that. The NAO is also zonally elongated, but the suggestion of wave energy passing through is weak. If one studies high-frequency filtered data, one is more likely to find transient teleconnections of the meridionally elongated variety, somewhat like the EWP1 dispersion of a source in mid-latitude. In reality, in unfiltered data, both types are present as can be seen from the EWP2 dispersion in Chapter 3. The reason EWP1 looks more like high frequency may be that the group speed is enhanced in the zonal direction by the background wind, see Appendix 2 of Chapter 3, so zonal dispersion is inherently on a faster time-scale than meridional dispersion.

While the PNA is likely explained in part by wave propagation,[7] the NAO does not fall in this category. The NAO remains somewhat of a mystery being highly weather related on the one hand (Franske et al. 2004) but often invoked to explain inter-decadal climate variability on the other (Hurrell 1995). Wallace and Thompson (1998) have speculated that the NAO is a manifestation of something more fundamental, namely variation in the zonal mean zonal wind, something relevant to all longitudes, not just the Atlantic basin. Indeed, in the stratosphere and the Southern Hemisphere such zonally invariant variations appear very important. In this context they introduced the "annular mode" [initially called the Arctic Oscillation (AO)], but the AO cannot be found in the Northern Hemisphere troposphere by traditional teleconnection methods (Ambaum et al. 2002; this chapter), although this point may be debatable (Wallace 2000). There is no counterpart to the NAO in the Pacific basin, at least nothing of that importance in terms of EV. Some studies have reported independent east and west Pacific Oscillations, but they are weak in EV. A fruitful approach is to study how NAO and PNA in their respective polarities change latitude and/or strength of the climatological jetstreams in the Pacific and Atlantic basin (Ambaum et al. 2002).

It would be an overstatement to say we understand teleconnections. Even if the PNA is explained by wave energy propagation, we have not explained why it is where it is, or why there are no PNA look-alikes at other longitudes. Moreover, no-one has ever seen the PNA or NAO; even at record breaking projections the flow across the Atlantic does not look all that much like the canonical NAO. Another research issue: What is the

---

[7] To actually go from a demonstration of EWP2 in Figure 3.2 to a steady state teleconnection pattern like the PNA one needs to do some work. Figure 3.2 shows only the beginning of a transient phenomenon. One needs to reinforce at each time step the initial source (and the latitude and size of the source are also important), and add some (far field) dissipation. A time mean may have to be taken to retain only steady state response.

relationship of the Icelandic Low and the Azores High (two climatological fixtures) to the two main poles (anomalies) of the NAO?

The reader should note that a systematic search for teleconnections on seasonal mean $Z500$ from 20°N to the pole in JFM yielded the NAO and PNA, so where exactly is the ENSO teleconnection? Originally it was thought that the PNA is the vehicle that brings the tropical ENSO into the mid-latitudes, but this view is no longer universally held. The EOT approach allows one to chose a time series from outside the domain of analysis, here 20°N–pole, for instance a time series that represents ENSO. Figure 4.2 may be seen in this light. Forcing an ENSO pattern as the first mode explains only 8% of the $Z500$ variance. The next EOT modes after the first forced ENSO mode is removed are the PNA and NAO with 3% and 1% EV less than before. This suggests that NAO and PNA in $Z500$ are modes primarily internal to mid-latitudes. Does this mean ENSO is unimportant? ENSO becomes a more important component by any of the following steps: (1) extend the domain slightly southward to 10°N or the equator, (2) standardize the $Z500$ variable, i.e. de-emphasize the high variance $Z500$ areas in high latitudes, or (3) use streamfunction instead of height. Some of the calculations in Chapter 5 will bring out this point.

4.6   Monitoring, indices and station data

Because of their importance, several institutions keep track of NAO, PNA, etc. in real time. The favored method is to express the state of these modes by an "index". The index could, in principle, be something mildly compli-cated (projection coefficients of an EOF), but is usually more simple. In view of Figure 4.1 and in the spirit of Wallace and Gutzler (1981) an NAO index could be defined as (a) the height anomaly at a single point 50°W, 65°N ($Z'(50,65)$), or (b) $(Z'(50,65)-Z'(50,30))/2$. Because these two locations are supposedly in a high negative correlation the average in option (b) is a less noisy estimate of the state of the NAO. (Because tradition says that high index corresponds to strong westerlies the NAO index is actually the negative of (a) and (b).) Some have taken areal averages around such points in order to further suppress noise. The PNA would be an average of four locations with the sign reversed between the even and odd entries. Standardization can be applied to force the index to have unit standard deviation. There are no official indices, accepted by a majority of researchers or institutions.

As long as we deal with gridded data one can chose the appropriate optimal grid points. But this is possible only in the post-1948 era. Extending the indices back to the early twentieth and nineteenth century is possible only by making a few compromises. The gist of the main compromise is to adjust the index to where we happen to have station data. Another is to use surface data only. For the NAO one still gets a good index, check Figure 4.1, by using one sea-level pressure station in say Iceland, and the other in Portugal. (This statement is mildly incorrect if the spatial pattern itself has secular changes or varies too much with season—the latter appears to be case for the NAO; see Portis et al. 2001.) This explains the sudden fame of humble places like Stykkishólmur in Iceland and Ponta Delgada on the Azores. Both of these stations have data back to the mid-nineteenth century. Lisbon and Gibraltar have been promoted as alternatives for the Azores.

Much the same can be said about the ENSO index. Gridded tropical analyses have not been available for more than 25 years, so station data had been used widely prior to 1980 to measure the so-called Southern Oscillation Index (SOI) as sea-level pressure at Tahiti minus Darwin in Australia (Chen 1982). Subsequent global Reanalyses have indicated somewhat better situated grid points but the 'equatorial SOI' (Kousky personal communication 1999) has not caught on. Moreover for extension back into the nineteenth century one must use Tahiti, Darwin (or Batavia/Djakarta). Meanwhile the atmospheric SOI is losing in popularity against an oceanic definition (called mysteriously "Nino3.4", see Barnston et al. 1997), which is the SST averaged over $5°S$–$5°N$ and $170°W$–$120°W$. For the PNA we have less luck in building longer data sets because there is a lack of surface station data in the Pacific, even today.

There is an apparent contradiction in using a single (or a few) points and measuring the state of a large-scale pattern spanning the Earth. This is the mystery of teleconnections. One can project data onto a large-scale pattern, but the time series of the projection coefficients is very highly correlated to the data at a single point (which was selected for having that property). It is truly remarkable that pressure at a single location like Darwin in the fall has useful predictive information for winter in some far away mid-latitude areas. All observations are irreplaceable, but some observations are even more valuable than others.

4.7   Closing comment

In this chapter we discussed *simultaneous* teleconnections, as is done in most teleconnection studies. In truth teleconnections are not simultaneous. If there is a perturbation somewhere in the system, it takes days or weeks or months for the effect to be felt far away. Moreover, the restriction of simultaneity would appear to reduce application to prediction. So why are (simultaneous) teleconnections so often mentioned in connection with seasonal prediction? The best illustration is ENSO. If we forecast a large positive or negative ENSO index (such as the Nino3.4 anomaly) for next winter we assume that the simultaneous teleconnection into the (lower) mid-latitudes will be automatically there. The interpretation of the simultaneous teleconnection is enriched by the interpretation of cause and effect, and how energy flows. The application of the teleconnection towards prediction is possible when the upstream cause is predictable to a certain degree. Application of diagnostic knowledge about the NAO and PNA towards prediction is much harder.

5 Empirical Orthogonal Functions

The purpose of this chapter is to discuss Empirical Orthogonal Functions (EOF), both in method and application. When dealing with teleconnections in the previous chapter we came very close to EOF, so it will be a natural extension of that theme. However, EOF opens the way to an alternative point of view about space–time relationships, especially correlation across distant times as in analogues.[1] EOFs have been treated in book-size texts, most recently in Jolliffe (2002), a principal older reference being Preisendorfer (1988). The subject is extremely interdisciplinary, and each field has its own nomenclature, habits and notation. Jolliffe's book is probably the best attempt to unify various fields. The term EOF appeared first in meteorology in Lorenz (1956). Zwiers and von Storch (1999) and Wilks (1995) devote lengthy single chapters to the topic.

Here we will only briefly treat EOF or PCA (Principal Component Analysis) as it is called in most fields. Specifically we discuss how to set up the covariance matrix, how to calculate the EOF, what are their properties, advantages, disadvantages etc. We will do this in both space–time set-ups already alluded to in Equations (2.14) and (2.14a). There are no concrete rules as to how one constructs the covariance matrix. Hence there are in the literature matrices based on correlation, based on covariance, etc. Here we follow the conventions laid out in Chapter 2. The post-processing and display conventions of EOFs can also be quite confusing. Examples will be shown, for both daily and seasonal mean data, for both the Northern and Southern Hemisphere. EOF may or may not look like teleconnections. Therefore, as a diagnostic tool, EOFs may not always allow the interpretation some would wish. This has led to many proposed "simplifications" of the EOFs, which hopefully are more like teleconnections.

However, regardless of physical interpretation, since EOFs are maximally efficient in retaining as much of the data set's information as possible for as few degrees of freedom as possible they are ideally suited for empirical modeling. Indeed EOFs are an extremely popular tool these days. A count in a recent issue of the *Journal of Climate* shows at least half the articles using

---

[1] A pair of analogues are two states in a geophysical system, widely separated in time, that are very close.

EOFs in some fashion. Moreover, EOFs have some unique mind-twisting properties, including bi-orthogonality. The reader may not be prepared for bi-orthogonality, given the name empirical orthogonal functions and even Jolliffe's working definition given below.

5.1 Methods and definitions

5.1.1 Working definition

Here we cite Jolliffe (2002, p. 1). "The central idea of principal component analysis (PCA) is to reduce the dimensionality of a data set consisting of a large number of interrelated variables, while retaining as much as possible of the variation present in the data set. This is achieved by transforming to a new set of variables, the principal components, which are uncorrelated, and which are ordered so that the first *few* retain most of the variation present in all of the original variables." The italics are Jolliffe's. PCA and EOF analysis is the same.

5.1.2 The covariance matrix

One might say we traditionally looked upon a data set $f(s, t)$ as a collection of time series of length n_t at each of n_s grid points. In Chapter 2 we described that after taking out a suitable mean { } from a data set $f(s, t)$, usually the space-dependent time-mean (or "climatology"), the covariance matrix Q can be formed with elements as given by (2.14):

$$q_{ij} = \sum_t f(s_i, t)\, f(s_j, t)/n_t$$

where s_i and s_j are the ith and jth point (grid point or station) in space. The matrix Q is square, has dimension n_s, is symmetric and consists of real numbers. The average of all q_{ii} (the main diagonal) equals the space time variance (STV) as given in (2.16). The elements of Q have great appeal to a meteorological audience. Figure 4.1 featured two columns of the correlation version of Q in map form, the NAO and PNA spatial patterns, while Namias (1981) published all columns of Q (for seasonal mean 700 mb height data) in map form in an atlas.

5.1.3 The alternative covariance matrix

One might say with equal justification that we look upon $f(s,t)$ alternatively as a collection of n_t maps of size n_s. The alternative covariance matrix Q^a contains the covariance in space between two times t_i and t_j given as in (2.14a):

$$q^a_{ij} = \sum_s f(s, t_i) f(s, t_j)/n_s$$

where the superscript "a" stands for alternative. Q^a is square, symmetric and consists of real numbers, but the dimension is $n_t \times n_t$, which frequently is much less than $n_s \times n_s$, the dimension of Q. As long as the same reference $\{f\}$ is removed from $f(s, t)$ the average of the q^a_{ii} over all i, i.e. the average of main diagonal elements of Q^a, equals the space–time variance given in (2.16). The average of the main diagonal elements of Q^a and Q are thus the same.

The elements of Q^a have apparently less appeal than those of Q (seen as PNA and NAO in Figure 4.1). It is only in such contexts as in "analogues", see Chapter 7, that the elements of Q^a have a clear interpretation. The q^a_{ij} describe how (dis)similar two maps at times t_i and t_j are.

When we talk throughout this text about reversing the role of time and space we mean using Q^a instead of Q. The use of Q is more standard for explanatory purposes in most textbooks, while the use of Q^a is more implicit, or altogether invisible. For understanding it is important to see the EOF process both ways.

5.1.4 The covariance matrix: context

The covariance matrix typically occurs in a multiple linear regression problem where $f(t, s)$ are the *predictors*, and any dummy *predictand* $y(t)$, $1 \le t \le n_t$, will do. Here we first follow Wilks (1995; p. 368–369). A "forecast" of y (denoted as y^*) is sought as follows:

$$y^*(t) = \sum_s f(s, t) \, b(s) + \text{constant}, \qquad (5.1)$$

where $b(s)$ is the set of weights to be determined. As long as the time mean of f was removed, the constant is zero.

The residual $U = \sum_t \{y(t) - y^*(t)\}^2$ needs to be minimized. $\partial U/\partial b(s) = 0$ leads to the "normal" equations, see Equation 9.23 in Wilks (1995), given by:

$$Q\,b = a,$$

where Q is the covariance matrix and a and b are vectors. The elements of vector a consist of $\sum_t f(t, s_i)y(t)/n_t$. Since Q and a are known, b can be solved for, in principle.

Note that Q is the same for any y. (Hence y is a "dummy".)

The above can be repeated alternatively for a dummy $y(s)$

$$y^*(s) = \sum_t f(s, t) \, b(t) \qquad (5.1a)$$

where the elements of b are a function of time. This leads straightforwardly to matrix Q^a. Equation (5.1a) will be the formal approach to constructed analogue, see Chapter 7.

$\mathbf{Q}$ and $\mathbf{Q}^a$ occur in a wide range of linear prediction problems and $\mathbf{Q}$ and $\mathbf{Q}^a$ depend only on $f(s, t)$, here designated as the predictor data set.

In the context of linear regression it is an advantage to have orthogonal predictors, because one can add one predictor after another and add information (variance) without overlap, i.e. new information not accounted for by other predictors. In such cases there is no need for backward/forward regression and one can reduce the total number of predictors in some rational way. We are thus interested in diagonalized versions of $\mathbf{Q}$ and $\mathbf{Q}^a$ (and the linear transforms of $f(s, t)$ underlying the diagonalized $\mathbf{Q}$'s).

5.1.5   EOF through eigenanalysis

In general a set of observed $f(s, t)$ are not orthogonal, i.e. $\Sigma f(s_i, t) f(s_j, t)/n_t$ and $\Sigma f(s, t_i) f(s, t_j)/n_s$ are not zero for $i \neq j$. Put another way: neither $\mathbf{Q}$ nor $\mathbf{Q}^a$ are diagonal. Here some basic linear algebra can be called upon to diagonalize these matrices and transform the $f(s,t)$ to become a set of uncorrelated or orthogonal predictors. For a square, symmetric and real matrix, like $\mathbf{Q}$ or $\mathbf{Q}^a$, this can be done easily, an important property of such matrices being that all eigenvalues are positive and the eigenvectors are orthogonal. The classical eigenproblem for matrix $\mathbf{B}$ can be stated:

$$\mathbf{B}\ e_m = \lambda_m e_m \qquad (5.2)$$

where e is the eigenvector and λ is the eigenvalue, and for this discussion $\mathbf{B}$ is either $\mathbf{Q}$ or $\mathbf{Q}^a$. The index m indicates there is a set of eigenvalues and vectors. Notice the non-uniqueness of (5.2): any multiplication of e_m by a positive or negative constant still satisfies (5.2). Often it will be convenient to assume that the norm $|e|$ is 1 for each m.

Any symmetric real matrix $\mathbf{B}$ has these properties:

(1) The e_m's are orthogonal.
(2) $\mathbf{E}^{-1}\ \mathbf{B}\ \mathbf{E}$, where the matrix $\mathbf{E}$ contains all e_m, results in a matrix $\mathbf{\Lambda}$ with the elements λ_m at the main diagonal, and all other elements zero. This is one obvious recipe to diagonalize $\mathbf{B}$ (but not the only recipe! see EOT in sct 5.3).
(3) All $\lambda_m > 0$, $m = 1, \ldots, M$.

Because of property (1) the $e_m(s)$ are a basis, orthogonal in space, which can be used to express:

$$f(s, t) = \sum_m \alpha_m(t) e_m(s) \qquad (5.3)$$

where the $\alpha_m(t)$ are calculated, or thought of, as projection coefficients, see Equation (2.6). But the $\alpha_m(t)$ are orthogonal by virtue of property #(2). It is actually only the second step/property that is needed to construct orthogonal predictors. (In the case of $\mathbf{Q}$, execution of step (2) implies that the time series $\alpha_m(t)$ (linear combinations of the original $f(s, t)$) become orthogonal and $\mathbf{E}^{-1}\,\mathbf{Q}\,\mathbf{E}$ diagonal.) Here we thus have the very remarkable property of bi-orthogonality of EOFs—both $\alpha_m(t)$ and $e_m(s)$ are an orthogonal set. With justification the $\alpha_m(t)$ can be looked upon as basis functions also, and (5.3) is satisfied when the $e_m(s)$ are calculated by projecting the data onto $\alpha_m(t)$.

We can diagonalize $\mathbf{Q}^a$ in the same way, by calculating its eigenvectors. Now the transformed maps (linear combinations of original maps) become orthogonal due to step (2), and the transformed time series are a basis because of property (1). (Notation may be a bit confusing here, since, except for constants, the e's will be α's and vice versa, when using $\mathbf{Q}^a$ instead of $\mathbf{Q}$.) One may write, as in Equation (5.2)

$$\mathbf{Q}\,\mathbf{e}_m = \lambda_m \mathbf{e}_m$$
$$\mathbf{Q}^a\,\mathbf{e}_m^a = \lambda_m^a \mathbf{e}_m^a \qquad (5.2a)$$

such that the e's are calculated as eigenvectors of $\mathbf{Q}^a$ or $\mathbf{Q}$. We then have, as in Equation (5.3)

$$f(s, t) = \Sigma \alpha_m(t) e_m(s)$$
$$f(s, t) = \Sigma e_m^a(t) \beta_m(s) \qquad (5.3a)$$

where the α's and β's are obtained by projection, and the e's are obtained as eigenvectors. When ordered by EV, $\lambda_m = \lambda_m^a$, and except for multiplicative constants $\beta_m(s) = e_m(s)$ and $\alpha_m(t) = e_m^a(t)$, so (5.3) alone suffices for EOF.

Note that $\alpha_m(t)$ and $e_m(s)$ cannot both be normed at the same time while satisfying (5.3). This causes considerably confusion. In fact all one can reasonably expect is:

$$f(s, t) = \Sigma \alpha_m(t)/c \times e_m(s)c \qquad (5.3b)$$

where $c\ (m)$ is a constant (positive or negative). Equation (5.3b) is consistent with both (5.3) and (5.3a). Neither the polarity, nor the norm is settled in an EOF procedure. The only unique parameter is λ_m.

Since there is only one set of bi-orthogonal functions, it follows that during the above procedure $\mathbf{Q}$ and $\mathbf{Q}^a$ are simultaneously diagonalized, one explicitly, the other implicitly for free. It is thus advantageous in terms of computing time to choose the covariance matrix with the smallest dimension. Often, in meteorology $n_t \ll n_s$. Savings in computer time can be enormous.

5.1.6   Explained variance (EV)

The eigenvalues can be ordered: $\lambda_1 > \lambda_2 > \lambda_3 > \ldots > \lambda_M > 0$. Moreover:

$$\sum_{m=1}^{M} \lambda_m = \sum_{i=1}^{n_s} q_{ii}/n_s = \text{STV}.$$

The λ_m are thus a spectrum, descending by construction, and the sum of the eigenvalues equals the space–time variance.

Likewise

$$\sum_{m=1}^{M} \lambda_m = \sum_{k=1}^{M} q^a_{kk}/n_t = \text{STV}.$$

The eigenvalues for $\mathbf{Q}$ and $\mathbf{Q}^a$ are the same. The total number of eigenvalues, M, is thus at most the smaller of n_s and n_t

In the context of $\mathbf{Q}$ one can also write:

explained variance of mode $m(\lambda_m) = \Sigma\alpha^2_m(t)/n_t$ as long as $|e| = 1$.

Jargon: mode m "explains" λ_m of STV or $\lambda_m/\Sigma\lambda_m \times 100\%$EV

The notion EV requires reflection. In normal regression one explains the variance of the predictand y. But here, in the EOF context, we appear to explain the variance in $f(s, t)$, the predictor. Indeed EOF is like self-prediction. Equation (5.1) is still true if the dummy $y(t)$ is actually taken to be the time series of $f(s, t)$ at the mth point: $f(s_m, t)$. Equation (5.1) then reads:

$$f^*(s_m, t) = \sum_s f(s, t)\, b(s, s_m)$$

where $b(s, s_m)$, for a fixed point s_m, is the regression coefficient between s_m and any $s(1 \leq s \leq n_s)$ and the EOFs are found by minimizing U

$$U = \sum_{m=1}^{n_s} \sum_t (f(s_m, t) - \sum_s f(s, t)\, b(s, s_m))^2. \tag{5.4}$$

Now $f(s, t)$ is both predictor and predictand. At this point the order $m=1, 2$, etc. is arbitrary. As per Equation (5.2) and executing $\mathbf{E}^{-1}\, \mathbf{B}\, \mathbf{E}$ the expression $\sum f(s, t)b(s, s_m)$ is transformed to $\Sigma\alpha_m(t)e_m(s)$.

For any truncation N ($1 \leq N < M \leq n_s$) the N functions retained are maximally efficient in EV because they minimize U.

5.2   Examples

Figure 5.1 gives an example of an EOF calculation. Shown are the first four EOFs following explicit diagonalization of $\mathbf{Q}$ by step (2), i.e. solving (5.2) for $\mathbf{B}=\mathbf{Q}$ and ordering by EV. The maps show $e_m(s)$ for $m = 1, \ldots, 4$; the time series underneath each map are $\alpha_m(t)$, $m = 1, \ldots, 4$ as per Equation (5.3), where $t=1948$–2005, 58 values at annual increment. The example corresponds exactly to the seasonal mean JFM 500 mb data on the $20°$N–pole domain already used in Chapter 4. The time mean removed is for 1971–2000, the current WMO climate normal. The time mean of each time series thus has zero mean over 1971–2000. The maps $e_m(s)$ are normed, see Appendix 1 for details, and the physical units are in the time

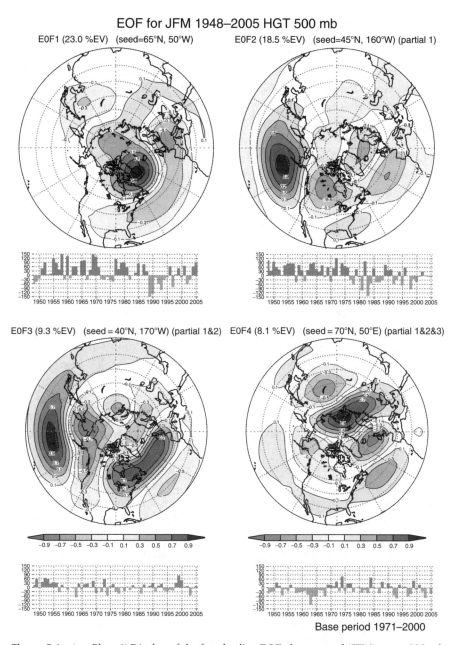

EOF for JFM 1948–2005 HGT 500 mb

EOF1 (23.0 %EV)   (seed=65°N, 50°W)   EOF2 (18.5 %EV)   (seed=45°N, 160°W) (partial 1)

EOF3 (9.3 %EV)   (seed = 40°N, 170°W) (partial 1&2)   EOF4 (8.1 %EV)   (seed = 70°N, 50°E) (partial 1&2&3)

Base period 1971–2000

Figure 5.1   (see Plate 4) Display of the four leading EOFs for seasonal (JFM) mean 500 mb height. Shown are the maps and the time series. A post-processing is applied, see Appendix 1, such that the physical units (gpm) are in the time series, and the maps have norm=1. Contours every 0.2, starting contours ± 0.1. Data source: NCEP Global Reanalysis. Period 1948–2005. Domain 20°N–90°N. The seed base point (e.g. 65°N,50°W) is mentioned because the iteration towards EOF (as described in Appendix 2) starts from the EOT.

series (gpm). One can thus see the amplitude of the time series going down with increasing m and decreasing EV. The EV of the first four EOFs is 23.0, 18.5, 9.3 and 8.1% respectively, for a total of nearly 60%. Indeed the first few modes explain a lot of the variance, as is the general idea of "principal components" analysis. EOF mode 1 and 2 still look like NAO and PNA, but explain somewhat more variance than the EOT counterparts in Figure 4.4. Moreover we did not have to identify some grid point as base point a priori as we had to for teleconnections and EOT.[2] While modes 3 and 4 explain much less variance than modes 1 and 2, they do have the attractive looks of dispersion on a sphere. The patterns shown are all orthogonal to each other (and to all 54 patterns not shown as well), but often in a complicated way (not like sin/cos). Likewise, the time series have zero inner product, but in a complicated way. For instance, the first two time series appear correlated by eye in the low frequencies, but this is compensated (and much harder to see for the human eye) by negative correlation on the inter-annual time-scale.

Figures 5.2 and 5.3 are EOFs for *daily* 500 mb data, for NH and SH, respectively, during 1998–2002 (DJF or 450 days). We use only 5 years here, but still have many more realizations than for seasonal means (58). The contrast between the two hemispheres could hardly be larger. In the NH, standing wave patterns looking like sweeping combinations of NAO and PNA dominate even daily data albeit at much reduced EV compared to seasonal means. In the SH EOFs on daily data suggest domination of surprisingly simple-harmonic-like wave motion in the west to east direction. This is certainly consistent with the relative success of EWP forecasts in the SH described in Chapter 3. It is amazing that so many variations of what looks basically like zonal wavenumber 4 can be spatially orthogonal. (For exactly four waves along 50°S there would be only two orthogonal arrangements). The time series in Figures 5.2 and 5.3 are for 450 daily points with zero mean and four discontinuities (at the end of February in 1998, 1999, 2000 and 2001). Even the daily time series show some low-frequency behavior with periods of 10–20 (or more) days of one polarity. Often the leading EOFs are thought of as displaying the large-scale *low-frequency* behavior, and most often they have been calculated from time-filtered data (e.g. monthly means) to force this to be the case. But this may not always be necessary; after all EOFs target the variance and given that the atmosphere has a red spectrum both in time and space it should be no surprise that low-frequency large-scale components show up first.

In Figures 5.2 and 5.3 the anomalies were formed first by subtracting an harmonically smoothed daily Z500 climatology based on 1979–1995

---

[2] However, note that we mention a base point (seed=65°N, 50°W) on top of each map. This is related to the initial guess used to start an iteration towards the EOFs, see Appendix 2. The initial guess is irrelevant, except for the polarity and the speed of convergence to EOF.

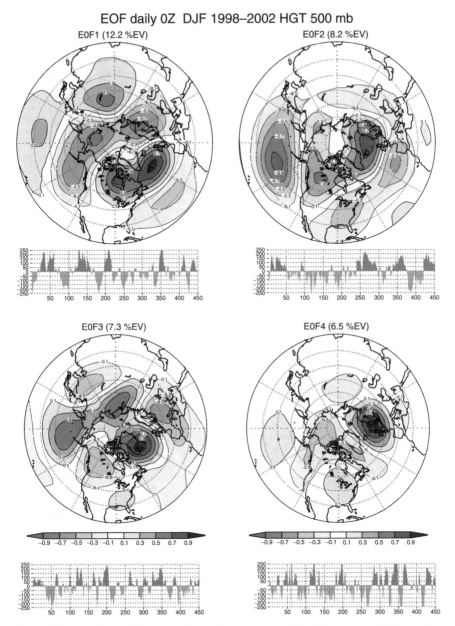

Figure 5.2   (see Plate 5) Same as Figure 5.1, but now daily 0Z data for all Decembers, Januaries and Februaries during 1998–2002

(Schemm et al. 1997). This leaves a considerable non-zero mean anomaly for a time series as short as 15 months (D, J or F) during 1998–2002. As a second step we subtracted the time mean across the 15 months to arrive at Figures 5.2 and 5.3. Without removing the time mean the first EOF in the SH is a (nearly) zonally invariant pattern with a nodal line at 55°S and a

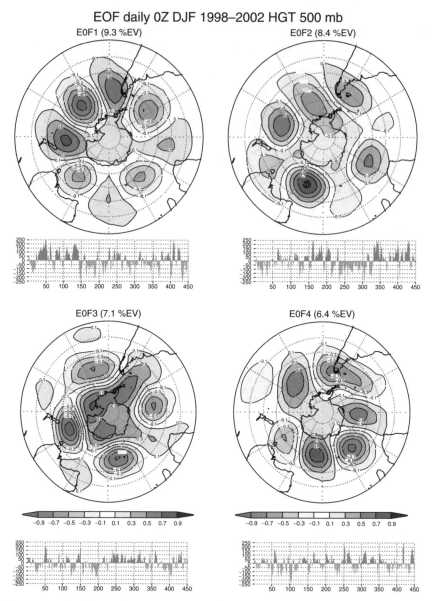

Figure 5.3   (see Plate 6) Same as Figure 5.2, but now the SH.

time series that is positive for most of the 5 years. Such "annular" variations are well established for the SH (Hartman 1995) and are known as the Southern Annular Mode or Antarctic Oscillation (AAO). Apparently the SH had stronger than average westerly mid-tropospheric flow in a band centered at 55°S during 1998–2002.[3]

---

[3] The name AAO was invented later as the counterpart for the AO (Thompson and Wallace 1998), which has, however, since been renamed Northern Annular Mode.

One should note that the EOFs (Figure 5.1) and the EOTs presented in Figure 4.4 are very similar. In this case EOF analysis does tell us something about teleconnections. The NAO and PNA renditions in Figure 4.4 are naturally orthogonal in space and very dominant in EV, so adding the constraint of spatial orthogonality will not change the results too much for these two modes. Usually the similarity between EOF and EOT is much less, and this could make interpretation of EOF (if one wants to see teleconnections) difficult. For instance in Figure 5.2 the leading EOFs are not readily seen as pure PNA and NAO, even though the EOT counterparts (not shown) would.

5.3   Simplification of EOF-EOT

Jolliffe (2002, p. 271 and Chapter 11) gives a list of methods to simplify EOFs, as practiced in various fields. The methods include "rotation", "regionalization", and EOT. Why the need for simplification? Often the blame is placed on the bi-orthogonality, which is allegedly too constraining or too mathematical to allow physical interpretation in many cases. In fact, this means that the way EOFs are calculated, as per Eq (5.2), is overkill. There is no need to calculate the eigenvalues of Q or Q^a if the only purpose is to create orthogonal predictors and to diagonalize either Q or Q^a (not both at the same time). All simplification of EOF methods appears to come down to relaxing orthogonality in one dimension (time or space) while maintaining orthogonality in the other. The EOT method, as described in Chapter 4, has orthogonal time series, thus diagonalizes Q, but the elements appearing on the diagonal are not the eigenvalues of Q. Notice that if the outer summation $m = 1, \ldots, n_s$ was not applied, a trivial solution presents itself for Equation (5.4): $b(s, s_m) = 1$ for $s = s_m$ and $b=0$ for all other s. This procedure yields the EOTs described in Chapter 4. The time series of the EOT attached to grid point s_m is $f(s_m, t)$ and thus trivially explains all variance at s_m. However, $f(s_m, t)$ explains far more variance of the whole field (STV) than it would if it just explained the variance at one point. This, of course, is because of the non-zero correlation between point m and most other points. The EOTs are not unique; one can start with any grid point and proceed with any choice for a next grid point.[4] Ordering m by EV makes sense and drives the EOT closer to EOF. Upon ordering, the matrix Q has been transformed to a diagonal matrix M (and thus orthogonal time series) with elements μ_m, where μ_m decreases with m. Although μ_m are not the eigenvalues of Q we still have $\Sigma \mu_m = $ STV. No spatial basis exists. So

---

[4] EOT is quite free, while EOF is completely constrained by bi-orthogonality. One can even choose time series from outside the domain of the data set of from another data set (as in Section 8.7) and start EOT analysis that way.

even though an equation like (5.3) is still valid, it is satisfied without the e's being orthogonal and the α's are not projection coefficient (i.e. should not and cannot be calculated by projection data onto the e's).

In the alternative case we have

$$U = \sum_{m=1}^{n_t} \sum_s (f(s, t_m) - \sum_t f(s, t)\, b(t, t_m))^2. \tag{5.4a}$$

In the same way as described before, (5.4a) diagonalizes Q^a to form M^a (and thus orthogonal maps, which are linear combinations of the original maps), and $\Sigma\mu_m^a = STV$. The time series are not orthogonal. On a mode by mode basis μ_m^a does not equal μ_m (in general). Although an equation like (5.3) is valid, the α's are not orthogonal.

The alternative EOTs for JFM Z500 mean over 20–90°N are shown in Figure 5.4. The alternative (or 'reverse') EOTs start with the observed field in 1989. At 18.3% EV this observed field explains more of the variance of the 58 fields combined than any other. After regressing 1989 out of the data set the "once reduced" observed field in 1955 emerges as the next leading EOT, etc. The normalization described in Appendix 1 was applied for display purposes, so we can compare Figure 5.4 to Figures 4.4 and 5.1 in that the spatial patterns have been forcibly normed. While the higher order modes all have a year assigned to them (like seed=1974), they look less and less like the observations in that year (because modes 1 to $m-1$ were regressed out). The alternative calculation has 50% EV for four modes and the results do not look much like Figure 4.4, but are sweeping large-scale fields nevertheless. The time series show a high positive value for the seed year. The leading alternative EOTs are "real" in the sense that they were actually observed at some point in time, a statement that cannot be made about the PNA, NAO (Figure 4.4) or EOFs. The utility of reverse EOT should become clearer in the chapter on (constructed) analogues.

Not only can regular EOT be calculated in any order for grid point m, one can also manipulate the results. If someone does not like a portion of EOT#1, for instance the NAO related covariance over east Asian in Figure 4.4, one can modify EOT#1 by blanking out this area (forcing zeros). This would be an example of surgical regionalization. After this start one can continue to find the second EOT, possibly blank out more areas, etc. This is the basis for Smith and Reynolds' (2003; 2004) Reanalysis for Sea-Level Pressure and SST for century-long periods. Smith and Reynolds also found by example that EOTs, for their SST/SLP data sets, are "nearly the same" as rotated EOF as arrived at by a popular recipe called "Kaiser varimax". The principal objective of rotation of EOFs is to obtain more regionalized "simple" structures that are also more robust to sampling uncertainty as mimicked by leaving one or a few years in or out of the data set. The maps in Figure 4.4 change little upon adding one year, while the maps in Figure 5.1 (certainly the modes 3 and 4) can change beyond recognition. Much work

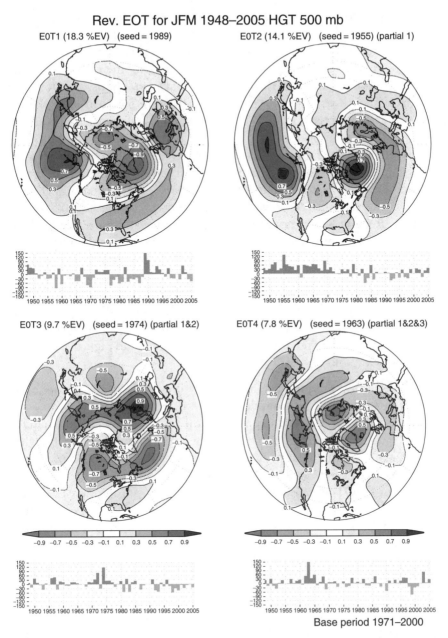

Figure 5.4 (see Plate 7) Display of four leading alternative EOTs for seasonal (JFM) mean 500 mb height. Shown are the regression coefficient between the base point in time (1989, etc.) and all other years (time series) and the maps of 500 mb height anomaly (geopotential meters) observed in 1989, 1955, etc. In the upper left for raw data, in the upper right map after removal of the first EOT mode, lower left after removal of the first two modes, etc. A post-processing is applied, see Appendix 1, such that the physical units (gpm) are in the time series, and the maps have norm=1. Contours every 0.2, starting contours ± 0.1. Positive values light shading, negative values darker shading. Negative contours are dashed. Data source: NCEP Global Reanalysis. Period 1948–2005. Domain 20°N–90°N

on rotation methods was done by Richman (1986), see also Richman and Lamb (1985). There are two types of rotation, leaving either the temporal or the spatial orthogonality intact (and sacrificing the other). There is an associated loss in EV for the leading modes, making the components less principal. The rotation performed commonly on atmospheric/oceanographic data is essentially to "rotate" from Figure 5.1 to roughly Figure 4.4, although not precisely so, relaxing orthogonality in space. The precise results of this rotation may be dependent upon the specified number of modes retained in the rotation (O'Lenic and Livezey 1988) and the rotation recipe. Since Figure 4.4 was obtained without setting the number of modes (in rotation), and even without a rotation recipe (like varimax) EOT seem an easy, natural and quick way to achieve many of the stated goals of rotation. Regular EOTs are more robust, simpler and more regionalized than EOFs. The alternative EOTs "regionalize" in the time domain, i.e. try to maximize projection on a flow that happened at a certain time.

Appendix 2 shows how EOFs can be calculated one by one by an iteration procedure, a good starting point for iteration are EOTs because they are already close to EOFs. Figure 5.5 shows the EOFs one obtains by starting from the alternative EOTs (and normalization as per Appendix 1). We obviously obtain the same EOFs as already shown in Figure 5.1, but we show them for a number of reasons:

(1) To show by example that there is only one set of EOFs, and one can iterate to those EOFs from drastically different directions.
(2) The polarity, while arbitrary, is set by the first guess. For example EOF1 in Figure 5.5 results from whatever the anomalies were in 1989, while in Figure 5.1 the polarity of EOF1 relates to the starting point being a base point near Greenland.
(3) The first EOT after EOF1 is removed differs from the first EOT after EOT1 is removed. While in Figure 5.4 1955 is the second year chosen, in Figure 5.5 it is 1948. Comparing Figure 5.1 to Figure 5.5 some modes come out in the same (opposite) polarity, but within the framework of (5.3a/b) these are obviously the same functions in both in time and space.

5.4   Discussion of EOF

5.4.1   Summary of procedures and properties

Figure 5.6 is an attempt to summarize in one schematic the various choices a researcher has and the operations one can perform. Obviously, one needs a data set, follow the two options (Q and Q^a), and see the three possibilities EOT, alternative EOT and EOF. In all three cases one can have a complete empirical orthogonal function representation of the data. Only in the latter

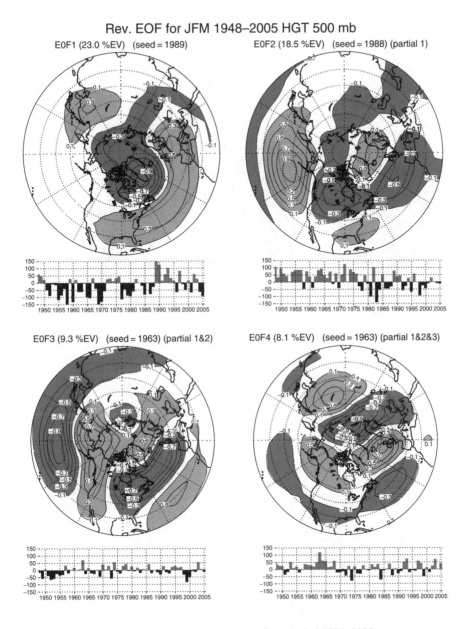

Figure 5.5 The same as Figure 5.1, but obtained by starting the iteration method (see Appendix 2) from alternative EOTs, instead of regular EOT. Compared to Figure 5.1 only the polarity may have changed. Positive values light shading, negative values darker shading. Negative contours are dashed.

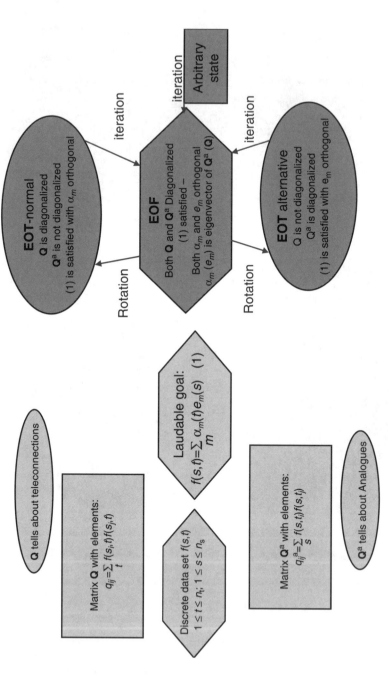

Figure 5.6 Summary of EOT/F procedures.

case is an eigenanalysis called for. The EOFs can be "rotated" in the direction of EOT. From any initial conditions one can iterate towards the gravest EOF.

5.4.2 The spectrum

Figure 5.7 shows the EV as a function of mode number for seasonal JFM $Z500$. For the EOF line this is also a spectrum of $\lambda(m)$. The EV by mode, left scale, decreases with m because of the ordering. The cumulative EV (right scale) obviously increases with m and is less noisy. The EOFs are more efficient than either version of EOT but never by more than a few percent (10% cumulative). The cumulative EV lines for EOT and EOF are best separated at about $m=8$ for this data set. But as we add more modes the advantage of EOF decreases. EOFs have a clear advantage only for a few modes. For either EOF or EOT only two modes stand out, or can be called principal components; already at mode 3 we see the beginning of a continuum of slowly decreasing poorly separated EV values. In this case the regular EOT are slightly more efficient than alternative EOT, but this is not generally true. This relative advantage depends on the number of points in time and space. If one imagines random processes taking place at a large number of grid points, the regular EOT would be very inefficient, on the order of EV $= 100\%/n_s$ per mode, while the alternative EOT cannot be less efficient than $100\%/n_t$.

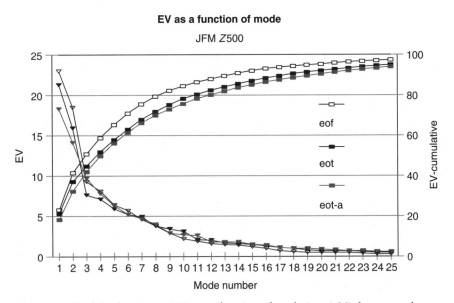

Figure 5.7 Explained variance (EV) as a function of mode ($m=1,25$) for seasonal mean (JFM) $Z500$, 20°N–90°N, 1948–2005. Shown are both EV(m) (scale on the left, triangles) and cumulative EV(m) (scale on the right, squares). White squares/triangles are for EOF, and black and grey for EOT and alternative EOT, respectively.

5.4.3   Interpretation of EOF

The interpretation of EOFs in physical terms is rarely straightforward. Adding to the difficulties are the vagaries of the procedure and display. The details of the covariance matrix, the exact domains used, the weighting of non-equal area grids, etc. varies. Even the display convention is confusing. Equation (5.3) consists of a time series with physical units multiplying a spatial field of non-dimensional regression coefficients. The latter are the spatial patterns of the EOF but many authors have displayed instead the correlation between the time series and the original data. While this may look better, these are not the EOFs. Nevertheless, in spite of these problems, IF (if!) there is an outstanding mode (like ENSO in global SST or the NAO in monthly or longer time mean sea-level pressure) any of the techniques mentioned will find them. Problems only arise with the less than principal modes, poorly separated from each other in EV, etc. On the other hand, why bother to interpret such modes?

5.4.4   Reproducibility (sampling variability)

We have described methods to calculate orthogonal functions from a given data set. To what extent are these the real EOFs of the population?. There are two different issues here. One is that the looks of a certain mode may change when the data set (a sample!) is changed slightly. Since EOF has been done often with a bias towards spatial maps and teleconnections this is a problem for the interpretation. Zwiers and von Storch (1999) are probably the best text in which sampling variability is addressed. The second important issue is the degree to which orthogonal functions explain variance on independent data. Even if modes are hard to reproduce they may continue to explain variance on independent data without too much loss, see examples in Van den Dool et al. (2000). Hence the sampling errors are mainly a problem for the physical interpretation, not for representing the data concisely. Here the North et al. (1982) rule regarding eigenvalues applies: Never truncate a λ_m spectrum in the middle of a shoulder (a shoulder or knee is an interruption in an otherwise regular decrease of eigenvalue with modenumber). Eigenvalues that are not well separated indicate possible problems with reproducibility (in terms of looks). Including all elements of the shoulder allows EV to be relatively unharmed on independent data. Recent work on this topic includes Quadrelli et al. (2005).

5.4.5   Variations on the EOF theme

In the literature one can find related procedures, such as "extended" EOFs (or "joint" or "combined" EOFs), abbreviated as EEOF. This means that two data sets $h(s, t)$ and $g(s, t)$ are merged into one $f(s, t)$, and then the EOF is done on $f(s, t)$ as before. For example Chang and Wallace (1987) combined a precipitation and temperature US data set and did EOF analysis on the combined data. It is convenient (but not necessary) to have the two variables on the same grid (stations). In that case EEOF is like extending the space domain to double the size. While EEOF is methodologically the same as EOF, one needs to make decisions about the relative weighting of the participating fields - in the case of precipitation and temperature one even has different physical units, so standardized anomalies may be an approach to place Temperature & Precipitation on an equal footing. If the number of grid points is different one could also adjust relative weighting to compensate. One has this problem already with EOF on a univariate height field from 20°N to 90°N. Because the standard deviation varies in space some researchers prefer standardized anomalies, for instance to give the tropics a better chance to participate (Barnston and Livezey 1987). This goes back to using covariance versus correlation matrices. There is no limit to the number of data sets in EEOF. Some authors merge time-lagged data sets to explore time-lagged (predictive) connections through EEOF. When two data sets are unrelated the EEOFs are the same as the original EOFs in the participating data, and chosen in turn. The first EEOFs in meteorology are probably in Kutzbach (1967).

Another procedure used is called *complex EOF*. This variation on EOF is used to specifically find propagating modes. In Chapter 3 we used sin/cos to diagnose propagation (and we could have called EWP "complex"). But EOFs are not analytical and not known ahead of time, so it is unclear whether any structures will emerge that are "90 degrees out of phase" so as to suggest propagation. Nevertheless, the EOFs based on daily data in the SH (Figure 5.3) certainly have the looks of propagating waves. Another way is to inspect the $\lambda(m)$ spectrum of regular EOF to search for pairs—in the case of propagation one expects two modes with nearly equal EV side by side. Such pairs (which may not be a case of poor separation) may indicate propagation, but this is not a sure sign. A more automated way of finding propagation is to transform $f(s, t)$ into $f(s, t) + \sqrt{(-1)} \times h(s, t)$, where $h(s, t)$ is the Hilbert transform of $f(s, t)$ (which is found by shifting all harmonics in $f(s, t)$ over 90 degrees). Complex EOF is done on transformed data. This method has been tried at least since the early 1980s; see for instance Horel (1984), Branstator (1987), Kushnir (1987) and Lanzante (1990).

5.4.6 EOF in models

If EOFs are so efficient, why haven't they replaced arbitrary functions like spherical harmonics in spectral models (very widely used in NWP from 1980 onward)? Essentially the use of EOF would allow the same EV for (far) fewer functions and be more economic in terms of CPU. EOFs are efficient in EV, but they are not efficient in CPU in transforming from a grid to coefficients and back. Spectral models did not become a success until the Fourier/Legendre transforms had been engineered to be very fast on a computer. Moreover, EOFs are not really functions, they are just a matrix of numbers, without easy (analytical) recipes for derivatives, interpolation, etc. A further drawback is that the EV efficiency of EOFs is about anomalies relative to a mean state, i.e. one would need to make an anomaly version of an NWP model to exploit the strength of EOF, not a very popular research direction either. Still EOFs have been investigated (Rinne and Karhilla 1975) and if CPU is not the issue they are obviously as good as any other function or better (Achatz and Opsteegh 2003). A final point about EV efficiency is that while EOFs are the most efficient, this feature in reality often applies only to a few functions that truly stand out (the principal components). After the first two modes are accounted for, the next functions all add very small amounts of variance, see Figure 5.7. If numerical prediction requires truncation at say 99% of the observed variance, EOFs are barely efficient relative to more arbitrary functions. Only when severe truncation to a few modes is required is EOF useful. Yet another issue is that accuracy in the time derivative (the main issue in prediction) may require insight in unresolved scales, or at least the interaction between the resolved and unresolved scales. This is somewhat possible with analytical functions, but is not easy with EOFs; see Selten (1993) who studied, as an alternative, the EOF of the time derivatives.

Other reasons to use observed EOFs in models may include a desire to study the energetics, stability, and maintenance of EOFs in a complete dynamical framework (Schubert 1985). This works only if the model's EOFs are close to observed EOFs.

Models formulated in terms of observed EOFs only use spatial orthogonality. No one has considered it possible to use the temporal orthogonality. It follows that one might as well use alternative EOTs. This would allow empirical access to the time derivative.

All of the above was about functions in the horizontal. EOFs are also efficient for representation of data in the vertical. Because there are fewer other (efficient or attractive) methods for the vertical, EOFs have been used in operational models for the vertical direction (Rukhovets 1963) for many years.

5.4.7 More examples

In this chapter we have shown just a few examples. It is beyond the scope of this book to investigate EOF/EOT of all variables at all levels throughout the seasonal cycle, both hemispheres, daily as well as time filtered data. The Barnston and Livezey (1987) study remains unique in that sense. Two final examples of EOF calculation will suffice here, see Figures 5.8 and 5.9. Figure 5.8 is an EOF analysis for global SST. The season chosen is OND, when the variance is near its seasonal peak, and the EV for the first mode is

Teleconnections JFM 1948–2005 HGT 500 mb

(bspnt = 65°N, 50°W)   (bspnt = 45°N, 160°W)

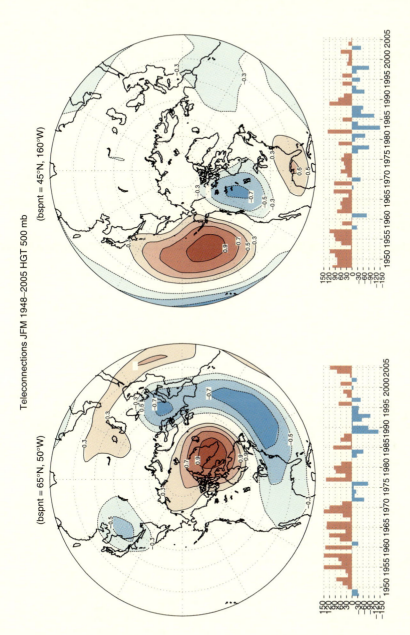

Plate 1 (corresponds to Fig 4.1): Display of teleconnection for seasonal (JFM) mean 500 mb height. Shown are the correlation between the base point (noted above the map) and all other grid points (maps) and the time series of 500 mb height anomaly (geopotential meters) at the base points. Contours every 0.2, starting contours ± 0.3. In both maps and time series the color red (blue) is used for positive (negative) values. Data source: NCEP Global Reanalysis. Period 1948–2005. Domain 20°N–90°N. On the left a pattern referred to as the North Atlantic Oscillation (NAO). On the right the Pacific North American pattern (PNA).

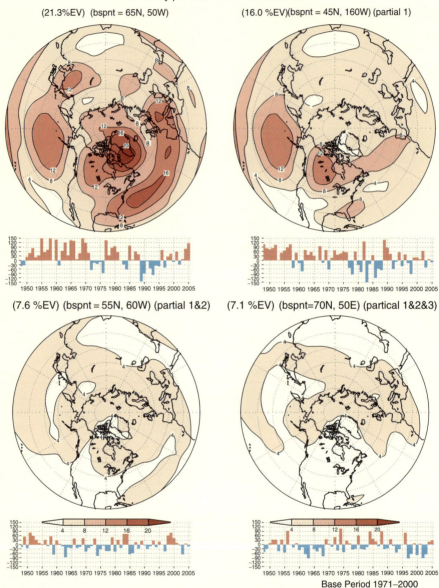

Domain EV by points: JFM 1948–2005 Z500

(21.3%EV) (bspnt = 65N, 50W)   (16.0 %EV)(bspnt = 45N, 160W) (partial 1)

(7.6 %EV) (bspnt = 55N, 60W) (partial 1&2)   (7.1 %EV) (bspnt=70N, 50E) (partical 1&2&3)

Base Period 1971–2000

Plate 2 (corresponds to Fig 4.3): EV(*i*), the domain variance explained by single grid points in % of the total variance of seasonal (JFM) mean Z500 over 1948–2005, using Equation (4.3). In the upper left for raw data, in the upper right after removal of the first EOT mode, lower left after removal of the first two modes, etc. Contours every 4%. Values in excess of 4% lightly shaded, in excess of 12% dark shading. The time series shown are the residual height anomaly at the grid point that explains the most of the remaining domain integrated variance.

normal EOT JFM 1948–2005 HGT 500 mb

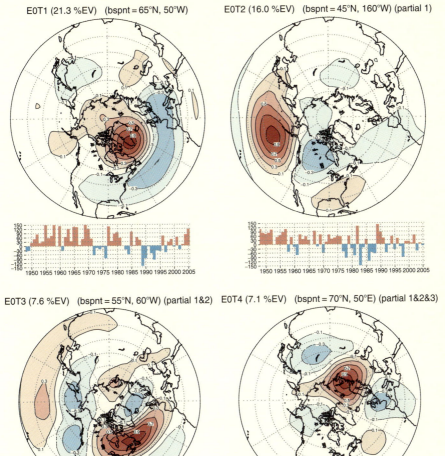

E0T1 (21.3 %EV) (bspnt = 65°N, 50°W)   E0T2 (16.0 %EV) (bspnt = 45°N, 160°W) (partial 1)

E0T3 (7.6 %EV) (bspnt = 55°N, 60°W) (partial 1&2)   E0T4 (7.1 %EV) (bspnt = 70°N, 50°E) (partial 1&2&3)

-0.9 -0.7 -0.5 -0.3 -0.1 0.1 0.3 0.5 0.7 0.9

Base period 1971–2000

Plate 3 (corresponds to Fig 4.4): Display of four leading EOT for seasonal (JFM) mean 500 mb height. Shown are the regression coefficient between the height at the base point and the height at all other grid points (maps) and the time series of the residual 500 mb height anomaly (geopotential meters) at the base points. In the upper left for raw data, in the upper right after removal of the first EOT mode, lower left after removal of the first two modes, etc. Contours every 0.2, starting contours ± 0.1. Data source: NCEP Global Reanalysis. Period 1948–2005. Domain 20°N–90°N. (A light post-processing was applied, See Appendix I of Chapter 5)

EOF for JFM 1948–2005 HGT 500 mb

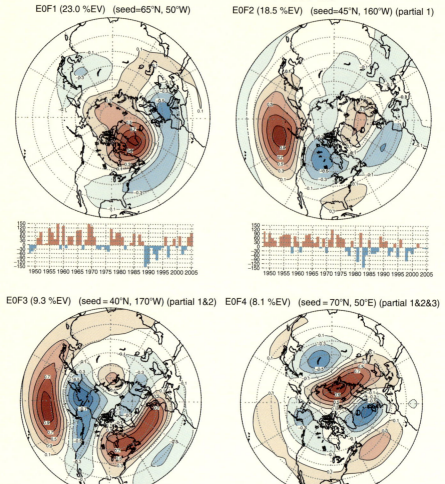

EOF1 (23.0 %EV)   (seed=65°N, 50°W)

EOF2 (18.5 %EV)   (seed=45°N, 160°W) (partial 1)

EOF3 (9.3 %EV)   (seed = 40°N, 170°W) (partial 1&2)

EOF4 (8.1 %EV)   (seed = 70°N, 50°E) (partial 1&2&3)

−0.9 −0.7 −0.5 −0.3 −0.1 0.1 0.3 0.5 0.7 0.9

Base period 1971–2000

Plate 4 (corresponds to Fig 5.1): Display of the four leading EOFs for seasonal (JFM) mean 500 mb height. Shown are the maps and the time series. A post-processing is applied, see Appendix 1, such that the physical units (gpm) are in the time series, and the maps have norm=1. Contours every 0.2, starting contours ± 0.1. Data source: NCEP Global Reanalysis. Period 1948–2005. Domain 20°N–90°N. The seed base point (e.g. 65°N,50°W) is mentioned because the iteration towards EOF (as described in Appendix 2) starts from the EOT.

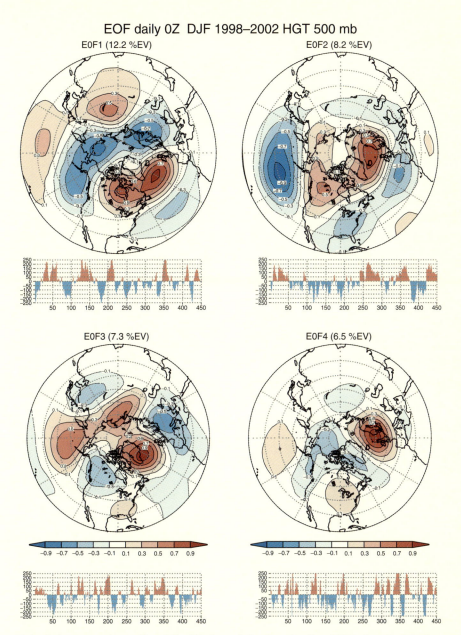

Plate 5 (corresponds to Fig. 5.2): Display of the four leading EOFs for the Northern Hemisphere based on <u>daily 0Z</u> 500 mb height data for all Decembers, Januaries and Februaries during 1998–2002. Shown are the maps and the time series. A postprocessing is applied, see Appendix I in Chapter 5, such that the physical units (gpm) are in the time series, and the maps have norm=1. Contours every 0.2, starting contours ± 0.1.

Data source: NCEP Global Reanalysis. Period 1948–2005. Domain 20N–90N.

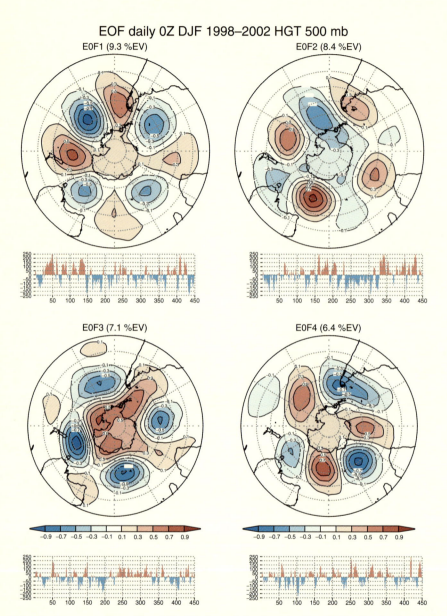

EOF daily 0Z DJF 1998–2002 HGT 500 mb

EOF1 (9.3 %EV)

EOF2 (8.4 %EV)

EOF3 (7.1 %EV)

EOF4 (6.4 %EV)

-0.9 -0.7 -0.5 -0.3 -0.1 0.1 0.3 0.5 0.7 0.9

Plate 6 (corresponds to Fig. 5.3): Display of the four leading EOFs for the Southern Hemisphere based on daily 0Z 500 mb height data for all Decembers, Januaries and Februaries during 1998–2002. Shown are the maps and the time series. A postprocessing is applied, see Appendix I in Chapter 5, such that the physical units (gpm) are in the time series, and the maps have norm=1. Contours every 0.2, starting contours ± 0.1.

Data source: NCEP Global Reanalysis. Period 1948–2005. Domain 20S–90S.

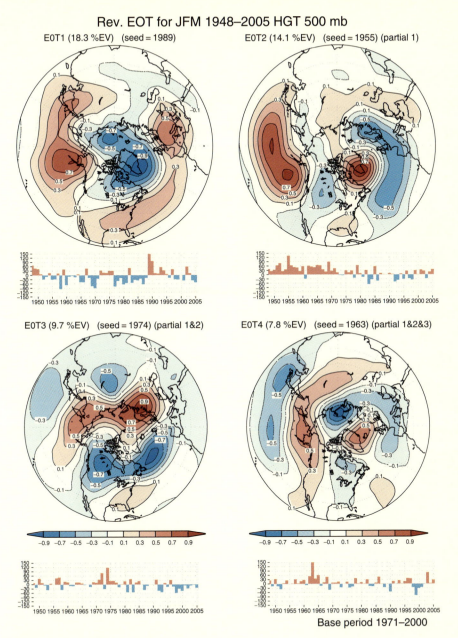

Rev. EOT for JFM 1948–2005 HGT 500 mb

E0T1 (18.3 %EV) (seed = 1989)

E0T2 (14.1 %EV) (seed = 1955) (partial 1)

E0T3 (9.7 %EV) (seed = 1974) (partial 1&2)

E0T4 (7.8 %EV) (seed = 1963) (partial 1&2&3)

Base period 1971–2000

Plate 7 (corresponds to Fig 5.4): Display of four leading alternative EOTs for seasonal (JFM) mean 500 mb height. Shown are the regression coefficient between the base point in time (1989, etc.) and all other years (time series) and the maps of 500 mb height anomaly (geopotential meters) observed in 1989, 1955, etc. In the upper left for raw data, in the upper right map after removal of the first EOT mode, lower left after removal of the first two modes, etc. A post-processing is applied, see Appendix 1, such that the physical units (gpm) are in the time series, and the maps have norm=1. Contours every 0.2, starting contours ±0.1. Positive values light shading, negative values darker shading. Negative contours are dashed. Data source: NCEP Global Reanalysis. Period 1948–2005. Domain 20°N–90°N

SST EOF OND 1955–2005

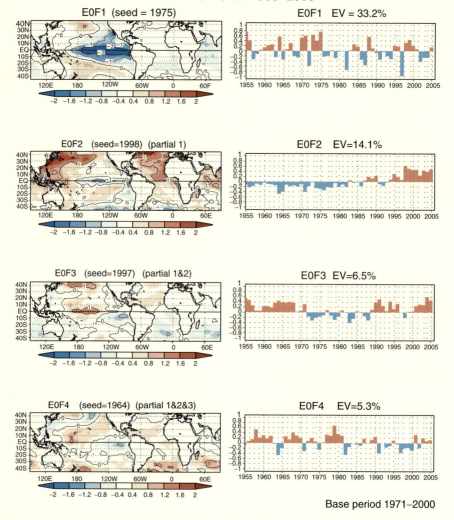

EOF1 (seed = 1975)   EOF1   EV = 33.2%

EOF2 (seed=1998) (partial 1)   EOF2   EV=14.1%

EOF3 (seed=1997) (partial 1&2)   EOF3   EV=6.5%

EOF4 (seed=1964) (partial 1&2&3)   EOF4   EV=5.3%

Base period 1971–2000

Plate 8 (corresponds to Fig 5.8): Display of four leading EOFs for seasonal (OND) mean SST. Shown are the maps on the left and the time series on the right. Contours every 1C, and a color scheme as indicated by the bar. Data source: NCEP Global Reanalysis. Period 1948–2005. Domain 45°S–45°N

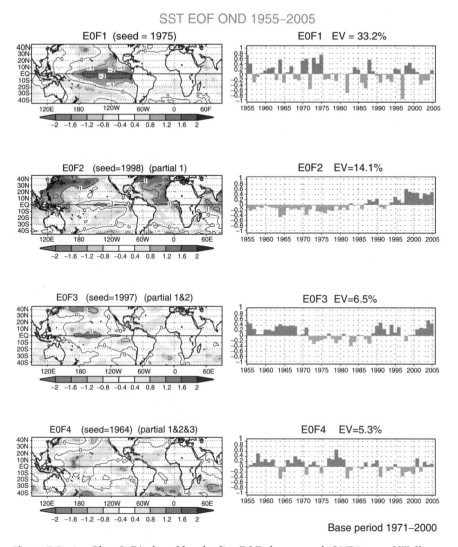

Figure 5.8   (see Plate 8) Display of four leading EOFs for seasonal (OND) mean SST. Shown are the maps on the left and the time series on the right. Contours every 1C, and a color scheme as indicated by the bar. Data source: NCEP Global Reanalysis. Period 1948–2005. Domain 45°S–45°N

well over 30%. The first mode absolutely stands out and it is identified as the ENSO mode as it manifests itself in global SST data. Starting the iteration from 1975, the polarity is the opposite of a warm event. The time series on the right indicates all famous warm event years (1972, 1982, 1997) as large negative excursions. Cold event years, especially 1998, are harder to trace in the time series as they project on more than one mode, i.e. cold events are not just the opposite of warm events. The second mode is very striking also and has a near uniform increase in the time series over 58 years, and indicates a

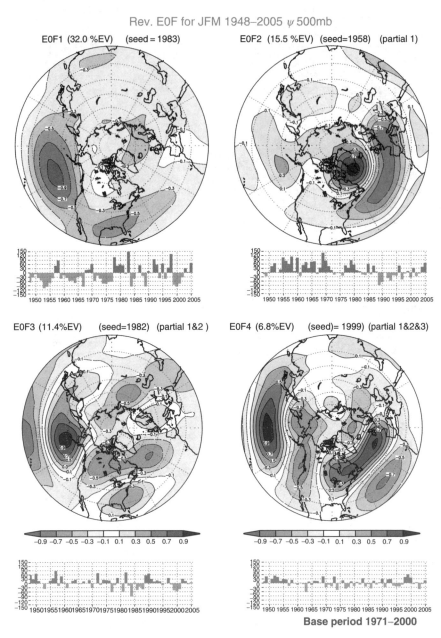

Figure 5.9 (see Plate 9) Display of four leading EOFs for seasonal (JFM) mean 500 mb streamfunction. Shown are the maps and the time series. A post-processing is applied, see Appendix 1, such that the physical units ($10^5 m^2/s$) are in the time series, and the maps have norm=1. Contours every 0.2, starting contours ± 0.1. Data source: NCEP Global Reanalysis. Period 1948–2005. Domain 20°N–90°N

pattern of warming SST in many oceans, except east of the date line along the equator in the Pacific and in parts of the Southern Oceans. This mode had its strongest positive projection in the 98/99 cold event. The third mode is one of inter-decadal variation, and a period of 30–35 years. At this point, however, modes explain only 5–6% of the variance.

Figure 5.9 gives more information about both sensitivity of EOF to details and the possible impact of ENSO on mid-latitude. Figure 5.9 is identically the same as Figure 5.5 except that we use the streamfunction (ψ) instead of height (Z). Since the horizontal derivatives of ψ and Z are approximations to the same observed wind these two variables are closely related in mid-latitude. Under geostrophic theory $\psi = Z/f$, where f is the Coriolis parameter. At lower latitude the variations in ψ are thus more pronounced than those in Z, although not nearly as much as when using the correlation instead of the covariance matrix for Z. Figure 5.9 shows the familiar NAO and PNA as modes 2 and 3 at somewhat reduced variance (3% and 2% less than in Figure 5.5, respectively), but they are preceded by a mode we have not seen before, moreover explaining 32% of the variance. One can tell from the high projection in 1983 (the seed) and 1998 that this pattern is active during warm ENSO years. The main action is a deep low in the North Pacific near 40°N and 160°W and a like signed anomaly over the US Gulf coast. Together these ψ centers modulate the subtropical jets in both oceans. While the first mode looks loosely speaking like a Pacific North American Pattern it is in fact orthogonal to the third pattern. Clearly, any statements on the impact of ENSO on the mid-latitudes, as per EOF analysis, requires careful study. Changing the domain size in Figure 5.5. down to 10°N, or the equator also has a large impact, because an ENSO mode would be first, followed by NAO and PNA.

5.4.8   Common misunderstandings

(a) *The first EOF "escapes" the drawbacks of being forced to be orthogonal because it is the first.* This opinion is wrong. There is no first EOF. All EOFs are simultaneously known by solving Equation (5.2), and this is true even if only one of them is calculated. Ordering of EOFs is entirely arbitrary relative to the calculation method. The opinion in italics is, however, correct for EOT. For instance in Figure 5.4 the field observed in 1989 is the first alternative EOT. Issues of orthogonality come in only when choosing the second EOT.

(b) *EOF patterns always show teleconnections.* This is wrong. They may or they may not. For instance, Figure 5.1 would suggest that the main center for the PNA in the Pacific is accompanied by a same signed anomaly in the Greenland/Iceland/Norway area. Figure 4.1 shows that the simultaneous correlations between the Pacific and the Atlantic are extremely weak. While the EOTs in Figure 4.4 are faithful to teleconnections (defined by linear simultaneous correlation), the EOFs in Figure 5.1 are not (everywhere).

(c) *EOF analysis forces warm and cold ENSO events to come out as each other's opposite.* While linearity has its drawbacks, this is not one of them. There is nothing against time series that have very strong values in a few years to be compensated by weak opposite anomalies in many other years; i.e. nothing in the procedure forces strong opposing values in just a few years. Figure 5.8 demonstrates this for OND SST. The largest Pacific warm events are clear in the first mode (1972, 1982, 1997), but 1998 (a strong cold event) does *Not* have a strong opposite projection in the first mode. In fact cold events can only be reconstructed using several modes, so the asymmetry noted by several authors (Hoerling et al. 1997) is not butchered by the EOF procedure.

(d) *EOFs describe stationary patterns only.* Geographically fixed patterns are no drawback in describing moving phenomena. A pair of EOFs, each stationary in its own right, but handing over amplitude from one to the other as time goes on, describe a moving system. Daily unfiltered data, full of moving weather systems, can be described 100% by EOFs.

5.4.9   Closing comment

We entered this chapter in classical fashion, the **Q** matrix, one-point teleconnections, regular EOT, and associated diagnostics but we want to leave it in the alternative fashion, emphasizing the $\mathbf{Q}^a$ matrix. The search for analogues (Chapter 7), the featured recipe to calculate degrees of freedom (Chapter 6), alternative EOT and the construction of analogues (Chapter 7) all use $\mathbf{Q}^a$. Linking orthogonal functions to specific moments in times past has one major advantage, i.e. direct access to the temporal evolution. If one expresses a specific state in the atmosphere as a linear combination of alternative EOTs one can easily make a forecast using a linear combination of the states that followed.

Appendix 1: Post processing (applies to Figures 4.4, 5.1, 5.2, 5.3, 5.4, 5.5 and 5.9)

On the graphical presentation of EOT and EOF, units, normalization, etc., let's assume that a data set $f(s, t)$ can be represented by (5.3) or (5.3a). We now discuss some post-processing, which in a nutshell is a matter of finding a factor c by which to divide e and multiply α. The left-hand side of (5.3) does not change in this operation:

$$f(s, t) = \sum_m \alpha_m(t)c(m) \, e_m(s)/c(m) \tag{5.3b}$$

$$f(s, t) = \sum_m \beta_m(s)/c^a(m) \, e^a_m(t)c^a(m). \tag{5.3c}$$

The factor c is a function of m, and c is different depending on whether we start with normal or reverse set-up, hence the superscript a. The reasons for doing these extra manipulations are varied. One could wish to have unit vectors in either space or time, regardless of how the calculations is done. Another reason is to make it more graphically obvious that EOFs obtained by normal and reverse calculation are indeed identically the same. Here we present a post-processing which makes, in all four (EOF/EOT, normal/reverse) possible cases, the spatial maps of unit norm, and places the variance and physical units in the time series.

Thus note the following:

(1) We plot maps and time series consistent with (5.3). We do not plot correlations on the map (as is very customary), because e is really a regression coefficient (when the α_m carry the physical units and the variance).

(2) To say that a mode m calculated under (5.3) is the same as mode m calculated under (5.3a) only means that $\alpha_m(t)e_m(s) = \beta_m(s)e_m{}^a(t)$; but $e_m(s)$ and $\alpha_m(s)$ are not the same, in general. But with the appropriate c and c^a applied $\alpha_m(t) \times c(m) = e_m{}^a(t) \times c^a(m)$ and $\beta_m(s)/c^a(m) = e_m(s)/c(m)$, except for sign.

(3) In order to force two identical modes to actually look identically the same, we do the following. We divide the map at each point by its spatial norm (and multiply the time series by this same norm so as to maintain the left-hand side as in (5.3b, 5.3c). The spatial norm of $e_m(s)$ is defined as

$$c(m) = \{\Sigma e_m(s)e_m(s)\}^{1/2},$$
$$c^a(m) = \{\Sigma \beta_m(s)\beta_m(s)\}^{1/2}$$

where the sum is over space. This action would make all maps of unit norm, and places the variance in the time series. The plots thus show, for example, $\alpha_m(t) \times c(m)$ as the time series.

(4) One can still tell how the calculation was performed, depending whether base points or seed years are mentioned in the label.

It turned out to be unsatisfactory vis-a-vis the contouring package to plot the normalized maps, so for cosmetics, we divide the map at each point by its absolute maximum value. This procedure creates maps with a maximum value of ± 1.

Appendix 2: Iteration

Often one needs to calculate only a few principal components. An insightful method is the so-called power method. Given Q, and an arbitrary initial state x_0, one simply repeatedly executes $\mathbf{x}_{k+1} = \mathbf{Q}\mathbf{x}_k$, $k = 0,1$, etc. In view of $Q\, e_m = \lambda_m e_m\,(5.2)$, x_{k+1} will converge to the eigenvector associated with the largest eigenvalue. This is so because if x_0 contains at least minimal projection onto e_1 that projection will be multiplied by λ_1 and since λ_1 is larger than all other eigenvalues the projection onto e_1 will ultimately dominate x_k for large k. Once x_k has converged the eigenvalue is found as the multiplication factor that results from executing $Q\, x_k$. At each step of the iteration one may need to set the norm of x_k equal to that of x_0 for stability. This methods only fails to find the first eigenvector if the separation of λ_1 and λ_2 is very very small, i.e. completely degenerate or pure propagation, a rare circumstance in practice. Once the first eigenvector is found one can proceed by removing the projection onto the estimate of e_1 from $f(s, t)$, then recalculate a new Q from once reduced $f(s, t)$. At this point the second eigenvalue should dominate, etc. Convergence is often very quick. An arbitrary guess x_0 will do but the EOTs are obviously a better initial guess. The iteration thus described is a rotation of the EOTs in the direction of EOFs. Van den Dool et al. (2000) does the iteration even without a covariance matrix.

6 Degrees of Freedom

How many degrees of freedom are evident in a physical process represented by $f(s, t)$? In some form questions about "degrees of freedom" (d.o.f.) are common in mathematics, physics, statistics, and geophysics. This would mean, for instance, in how many independent directions a weight suspended from the ceiling could move. Dofs are important for three reasons that will become apparent in the remaining chapters. First, dofs are critically important in understanding why natural analogues can (or cannot) be applied as a forecast method in a particular problem (Chapter 7). Secondly, understanding dofs leads to ideas about truncating data sets efficiently, which is very important for just about any empirical prediction method (Chapters 7 and 8). Lastly, the number of dofs retained is one aspect that has a bearing on how nonlinear prediction methods can be (Chapter 10).

In view of Chapter 5 one might think that the total number of orthogonal directions required to reproduce a data set is the dof. However, this is impractical as the dimension would increase (to infinity) with ever denser and slightly imperfect observations. Rather we need a measure that takes into account the amount of variance represented by each orthogonal direction, because some directions are more important than others. This allows truncation in EOF space without lowering the "effective" dof very much.

We here think schematically of the total atmospheric or oceanic variance about the mean state as being made up by N *equal* additive variance processes. N can be thought of as the dimension of a phase space in which the atmospheric state at one moment in time is a point. This point moves around over time in the N-dimensional phase space. The climatology is the origin of the phase space. The trajectory of a sequence of atmospheric states is thus a complicated Lissajous figure in N dimensions, where, importantly, the range of the excursions in each of the N dimensions is the same in the long run. The phase space is a hypersphere with an equal probability radius in all N directions. A similar plot in M dimensions ($M \gg N$), where each direction represents one EOF, would have the largest excursions for the gravest EOFs, and tiny excursions in a very large number of directions representing the higher EOFs, with an obvious truncation problem, M going to infinity as resolution goes to infinity. In a sense, N results from folding the energy in modes $> N$ back into modes $< N$, which serves to make N finite as well as creating a white spectrum in which the retained modes count as exactly one degree of freedom each. This makes the N processes somewhat hypothetical, a summary of the whole spectrum of variations in a single number: *the effective degrees of freedom*. As we shall see N, consistent with Toth (1995), is on the order of 30–50 for hemispheric daily instantaneous 500 mb height. Is 30 large or small? Thirty effective degrees of freedom is very large, as it renders the search for natural analogues virtually hopeless (van den Dool 1994). On the other hand 30 may seem very small compared to the nominal degrees of freedom (millions) used in NWP global models (the number of variables times the number of grid points); certainly we have a lot more spatial and/or temporal coherence than random processes at each point in time and space would have.

6.1   Methods to estimate effective degrees of freedom, N

Methods of estimating N have been reported infrequently in the literature. Moreover, the context and methods differ from paper to paper. Panofsky and Brier (1968) report that when correlating two time series of length N (drawn at random from the same parent distribution) the distribution of the correlation coefficient between the two time series is Gaussian with a zero mean and a standard deviation of $1/\sqrt{N-2}$, where N is the number of independent data in the time series (N is less than the length of the time series if autocorrelation were positive). Van den Dool (1981) extended this concept to maps (a string of data of length M) in order to test for statistical significance the spatial correlation between two anomaly maps (elements of the correlation rendition of $\mathbf{Q}^a$). It was noted that N in this case is the number of processes going on independently in space, i.e. much less than the number of grid points M because of correlation in space. If one correlates anomaly maps that should not be related (by virtue of them being years apart) one can retrieve N from an empirical pdf of correlations by inverting the expression $\mathrm{sd}_{cor} = 1/\sqrt{N-2}$. Following this approach we ask Nature to conduct its own Monte Carlo experiment. Livezey and Chen (1983) had a similar purpose in mind when designing their famous field significance test. They integrated the spatial correlation (elements of $\mathbf{Q}$) in space in order to find the integral space scale (S). S relates to N roughly as $N \approx$ domain size$/S$. This method is common in fluid dynamics and turbulence. Both Van den Dool (1981) and Livezey and Chen (1983) were concerned with the validity of claims of forecast skill by various methods. If someone's forecast anomaly map correlates 0.4 to the observed anomaly map, does that imply the forecast method is better than a random guess? Given N we can at least perform a statistical significance test.

Lorenz (1969) considered mean square differences (msd) between states in the atmosphere and, as explained in Toth (1995), a collection of msd should obey the chi-squared distribution with N degrees of freedom. From an empirical collection of msd one thus needs to find in some way the chi-squared pdf that gives the best fit, then retrieve N by an "inverse" method.

All these types of estimates of N (via msd, correlation), which work out numerically similar, but not identically the same, are closely related to classical notions of effective sample size, be it in time or space, or the de-correlation distance. Wang and Shen (1999) describe a more complete comparison of such methods, including theoretical background, and numerical estimates of N by each method on a given data set. Other aspects can be found in Fraedrich et al. (1995) and Stephenson (1997).

6.2 Example

Here we report on a fresh example using the global Reanalysis data for daily Z500 data at 0Z on the domain 20 degrees latitude to the pole for both hemispheres, as well as for the tropics between 20°N and 20°S. These are large areas in terms of km². We consider each calendar month separately. We first take out a smooth daily Z500 climatology, forming anomalies. We correlate all fields in, say, January, to all other fields in January but only in non-matching years. Therefore, the expected value of the covariance defined in Equation (2.14a) as

$$q^a_{ij} = \sum_s f(s, t_i) f(s, t_j) / n_s$$

is zero. q^a_{ij} are elements of the covariance matrix $\mathbf{Q}^a$, but we consider only those elements that have a large separation in t_i and t_j. Due to sampling q^a_{ij} is not always exactly zero but has a spread measured by its standard deviation. In terms of correlation defined as per Equation (4.1) as:

$$\rho_{ij} = q^a_{ij} / \sqrt{q^a_{ii} q^a_{jj}},$$

the expected value is 0, and the spread of ρ_{ij}, as per Gaussian theory (Panofsky and Brier 1968), is $1/\sqrt{N-2}$, where N is the effective degrees of freedom in the field f. If all M grid points had statistically independent equal variance processes going on, N would be M. But because of spatial correlation in the field $f(s, t)$ the value of N is (much) less than M.

From the collection of the K correlations (K is very large, about a million for 30 years of data) we calculate the standard deviation of the empirical correlation distribution (ECD) as

$$sd_{cor} = \{\Sigma \rho_{ij}^2/K\}^{1/2} \qquad (6.1)$$

where summation is over all admitted *ij* pairs and determine N as:

$$N = 1/sd_{cor}^2 + 2. \qquad (6.2)$$

This procedure makes sense only for $N > 3$, and generally only for large N. For small N, the ECD would feel the boundaries at $+1$ and -1.

Figure 6.1 shows how N depends on season in the two hemispheres. Obviously N is more or less proportional to the size (in km^2) of the domain (20 to the pole), but because weather systems have different horizontal length-scales in the two hemispheres and in winter versus summer there is clear variation in N. As shown by the lower curves in Figure 6.1, N in the NH varies from about 30 in winter to 45–50 in summer. In the SH, N is relatively constant at about 30 with little or no seasonality (presumably because the oceans are a very large fraction of the SH). Except for January the NH always has higher N than the SH. In summer, weather systems in the NH are indeed smaller than in winter.

The upper set of curves in Figure 6.1 are the radii of the N-dimensional hypersphere given by

$$sd = (\sum_{s,t} f(s, t)\, f(s, t)/n_s \times n_t)^{1/2}. \qquad (6.3)$$

This radius or climatological standard deviation is over 100 gpm in winter (in the respective hemispheres) but is as low as 65 in summer in the NH (a lesser factor that also contributes to a high estimate of N given imperfect

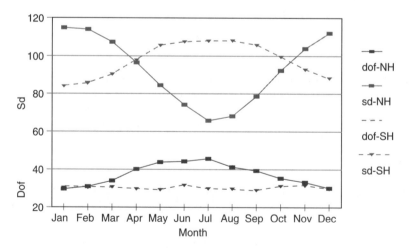

Figure 6.1   The seasonal cycle of N, the effective degrees of freedom (d.o.f; non-dimensional), and the standard-deviation (sd; gpm) for 500 mb daily height for the NH (20°N–pole) and the SH (20°S–pole). Period analyzed is 1968–2004. The lines for NH are full lines connecting squares, for SH dashed lines connecting triangles

analyses; summer has less of a signal). The SH during summer maintains a higher standard deviation of at least 85 gpm.

For the tropics (TR, not shown), the domain within 20 degrees from the equator, N is around 30 also, with small irregular seasonality. So we have three domains (NH, SH, TR) with similar N, except the NH summer which has higher N. The sd according to (6.3) is only 20 gpm for TR.

An estimate of N for three dimensions, all variables combined, appears to be of order 1000 (Toth 1995).

6.3   Link of degrees of freedom to EOF

Because EOFs do not have equal variance one may wonder how N relates to EOFs. It has been suggested recently that N can also be obtained through an integral involving the decrease of variance with EOF mode number (Bretherton et al. 1999):

$$N = \frac{(\sum_k \lambda_k)^2}{\sum_k \lambda_k^2} \qquad (6.4)$$

where λ_k is the variance explained by the kth ordered EOF, or the kth eigenvalue of the covariance matrix. For most typical decays of eigenvalue with mode, the value of N equals approximately the number of EOFs needed to describe 90% of the variance. So, as a rule of thumb, the value of N relates approximately to a tick mark on the cumulative variance versus EOF mode number graph (Huang et al. 1996). For instance, Figure 5.7 is such a graph and it would appear that seasonal mean Z500 in JFM (NH) have about 20 degrees of freedom. This number is lower than the $N \sim 30$ in Figure 6.1 because of temporal averaging (3 month means), which filters out many smaller scale and high-frequency weather systems and so decreases N.

For large N, Equations (6.4) and (6.2) give very similar numerical answers, with (6.2) being far easier to execute, so we used Equation (6.2) in the above. However, for interpretation's sake, Equation (6.4) is more helpful, especially since it shows the relationship to EOFs. For instance, if there was only one EOF N should be 1 according to (6.4), a most reasonable result. If the atmospheric variability is represented by K equal variance EOFs, (6.4) reduces to $N=K$, which is exactly the interpretation of N we are seeking.

Note that (6.4) is more generally valid than (6.2) because it also applies for low values of N. Recently (6.4) has been independently proposed by Patil et al. (2001) in the context of estimating the (sometimes low) dimension of an ensemble of forecasts, the name given by Patil et al. being BV- or E-dimension.

Comparisons of N in atmospheric analyses and model generated data have been made since Van den Dool and Chervin (1986); there are discrepancies, especially as the forecast lead increases, but there is broad agreement, even in the lower resolution models used in the 1980s. We made a recent calculation of N (for $Z500$ 20–pole) from 5 years of "reforecast" data (Schemm, personal communication) in NH winter. Figure 6.2 shows that despite a few curious ups and downs, N is very nearly correct out to 30 days. This does not prove that each of the 30 or so independent processes is identically the same as in Nature, although study of rotated EOFs appears to indicate mode-by-mode similarity for at least the first six EOFs (Peng personal communication). In general, modern day atmospheric models appear to be close copies of Nature on the resolved scales. In addition to N, the standard deviation around the mean, see Equation(6.3), should also be maintained, see the dashed line (labeled AMP) in Figure 6.2. Here, too, models are quite close to Nature. Perhaps 30 days is not quite enough to arrive in the model's true climate, and the small ups and downs in Figure 6.2 are part of transitioning from Nature to model climate.

6.4 Remaining questions

Some methods work with correlation in time, others with correlation in space. Possibly there are two values of N, depending on whether one has the regular or the reverse space–time point of view (of the same data set), in the

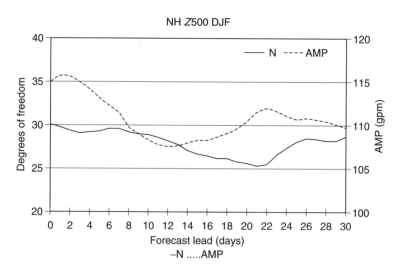

Figure 6.2 The dependence of the degrees of freedom (N), full line, on the lead of the forecast in a 5 year retrospective forecast data set with a T62L28 NCEP model of vintage 2002. The dashed line is the standard deviation around the model's climate mean, see Eq (6.3). Domain 20°N–north pole.

same way we have EOT and alternative EOT. The number of spatially independent processes does not (?) need to be the same as the number of temporally independent processes.

Another puzzling feature is that (6.2) is evaluated by correlating fields that, by construction, have an expected correlation 0 (by being far removed in time), while EOF calculations, followed by (6.4), are done inclusive of correlation between neighbors in time. By including neighbors in time (today and tomorrow) the ECD would have a secondary maximum at $+0.7$ for the daily 500 mb height anomaly due to persistence, and N can no longer be retrieved as per (6.2) because the Gaussian assumption is violated. A complimentary thought is as to how EOFs would be calculated if one deleted the correlation to neighbors in time or space from the covariance matrix; with EOT this might be a doable exercise. This puzzle is a companion to the distinction between Wallace and Gutzler (1981) type teleconnection (high correlation at remote distance) as opposed to EOF and EOT which place a premium on explaining variance both nearby (the bulk) and far away.

A third issue is that in most modern texts correlations would be z-transformed, i.e. $z = 0.5\log[(1 + r)/(1 - r)]$, where r is sample correlation, so the ECD based on z is more strictly Gaussian. In that case $sd_Z = 1/\sqrt{N - 3}$ or even $sd_Z = 1/\sqrt{N - 4}$ (Wang and Shen 1999). We believe we avoided the need for this complication by setting up Nature's Monte Carlo test, such that the expected correlation is zero, and the resulting ECD can be symmetric around zero. The Z transform is needed when the expected value is non-zero (and the ECD asymmetric).

A final issue to be mentioned is that N as derived here is for a single data set. There are applications when one wants to know how many d.o.f. two data sets have in common. Some of the references (Bretherton et al. 1999) describe some attempts, but much more work is needed.

6.5   Concluding comments

A value for N of about 30 implies that atmospheric states picked at random rarely (5% of the time) correlate more than ±0.38. This presents a bleak prospect for finding natural analogues or anti-analogues deserving of that name. Chances are a little better in the SH than in the NH, and especially poor in the NH during its summer. In Chapter 7 we will inspect the wings of the ECD to make further statements about naturally occurring analogues.

7 Analogues

In 1999 the Earth's atmosphere was gearing up for a special event. Towards the end of July, the 500 mb flow in the extra-tropical SH started to look more and more like a flow pattern observed some 22 years earlier in May 1977. Two trajectories in the N-dimensional phase space, N as defined in Chapter 6, were coming closer together. Figure 7.1 shows the two states at the moment of closest encounter, with the appropriate climatology subtracted. These two states are, for a domain of this size, the most similar looking patterns in recorded history.[1] But are these good analogues? They do look alike, nearly every anomaly center has its counterpart, but they are certainly not close enough to be indistinguishable within observational error, the anomaly correlation being only 0.81. The rms difference between the two states in Figure 7.1 is 71.6 gpm, far above observational error

500 mb height anomaly

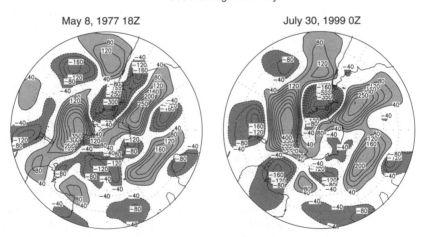

Figure 7.1 The most similar looking 500 mb flow patterns in recorded history on a domain this size (20° to the pole). These particular analogues were found for the SH, 20°S–90°S, and correlate at 0.81. The climatology, appropriate for the date and the hour of the day, was removed. Contours every 40 gpm. Positive values light shading, negative values darker shading. Negative contours are dashed. No zero line shown.

---

[1] Slight exaggeration. We looked at data for 1968–2004 only.

(<10 gpm). The close encounter did not make it to the newspapers and, more telling, not even to a meteorological journal.

The idea of situations in geophysical flow that are analogues to each other has always had tremendous appeal, at least in meteorology. Even lay people may comment that the weather today or this season reminds them of the weather in some year past. The implications of true analogues would be enormous. If two states many years apart were nearly identical in all variables on the whole domain (of presumed relevance), including boundary conditions, then their subsequent behavior should be similar for some time to come.[2] In fact one could make forecasts that way, if only it was easy to find analogues from a "large enough" data set. The analogue method was fairly widely used for weather forecasting at one time (Schuurmans 1973) but currently is rarely used for forecasts *per sé* (for all the reasons explained in Section 7.1). Rather analogues are used to specify one field given another, a process called "specification" or downscaling, or to learn about predictability (Lorenz 1969). In Section 7.1 we review the idea and limitations of naturally occurring analogues, and explain why/when it is (un)likely to find analogues. Finding analogues, as such, is a diagnostic problem. If no analogues deserving of that name exist in a finite data set, the application towards forecasting does not even arise. Section 7.2 develops the idea of "constructing" an analogue in the absence of any natural analogues (NA). The constructed analogue overcomes the main problem one has with natural analogues, although at some cost, and appears to have forecast applications (7.4). In the process of constructing an analogue, we make an empirical operator which allows us to address the calculation of unstable modes from just observations (7.6), study weakly nonlinear processes, and the dispersion of initial sources (much like in Figures 3.1 and 3.2), but in a more realistic way than EWP.

7.1   Natural analogues (NA)

The working definition for the existence of natural analogues is that two states in a dynamical system are so close they can be called each other's analogue. One qualifier is that these two states should be far apart in time, well beyond the de-correlation time. Two successive states could obviously be very close to each other, especially when observed at high temporal resolution, but the application to forecasting of such look-alike temporal neighbors would be meaningless as they stay temporal neighbors forever. We mean two states far apart in time, which by sheer coincidence[3] happen

---

[2] One should not expect exact analogues, because exact analogues would imply a periodic system, hence perfect predictability. Rather, we have states in mind that are close initially, with increasing differences as time progresses.

[3] One may always wonder about coincidence or chance, or the alternative that something forces the flow to be similar on specifically those dates. We remove the daily and annual cycle from the data and, in doing so, remove the effect of the main periodic forcings known to us.

to be close. We are interested in "how close" two states can be on a given data set of size M, how long they will stay close, how long they have been close (i.e. the ramping up to the closest encounter; symmetry in time or not), etc. and if we can't find any worthwhile analogues, why not? We are also interested in the most dissimilar flow patterns, the highest negative correlation, as it throws light on issues of linearity and symmetry. Importantly, we remove periodic components from the system (daily and annual cycle) before searching for analogues on the anomalies.

The age old idea of a forecast based on analogues is displayed in Figure 7.2. Today's point in phase space (or alternatively: today's weather map) is called the "base", and has time $t=0$ assigned to it. Then our task is to make a forecast for what will happen next. We look for analogues, i.e. cases in the past[4] that are very close to the base. These cases have to be around the same time of the year or at least subject to the same general climatic conditions. Having found the appropriate cases we line up the time axes and assign $t=0$ to these analogues also. The string of realizations, or the trajectory in phase space observed in the analogue year, following the analogue at $t=0$ is the forecast for the conditions that follow the base. If one compares this process to NWP, one might look upon the analogue as an analysis of the base, with an initial difference or error, and nature itself as the model that carries out the integration of the equations. The advantage of analogues, apart from simplicity, would be the use of a perfect model.[5] The problem with analogues is that the initial difference cannot be made small unless we either have an inordinate amount of data or very few degrees of freedom.

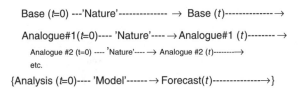

Figure 7.2 The idea of analogues. For a given "base", which could be today's weather map, we seek, in an historical data set, an analogue in roughly the same time of the year. The base and analogue are assigned $t=0$. The string of weather maps following the analogue would be the forecast for what will follow the base. (Data permitting there could be more than one analogue.) As a process this is comparable to "integrating" the model equations starting from an analysis at $t=0$, an analysis which is as close as possible to the true base. "Nature" stands for a perfect model.

---

[4] There is no difference in the role of base and analogue. If a state at time t_1 is very close to a state at t_2, the reverse is also true. It is only in real time that we cannot search for analogues in the future.

[5] Assuming the atmosphere–ocean–land system has not changed over time. The continual change in atmospheric composition, change in land use, etc. does violate this assumption, but we ignore this complication.

7.1.1   Similarity measures

The measure of similarity or analogy for two anomaly "maps" observed at t_i and t_j includes any of the following three expressions:

(a) RMSD, or root mean square difference:

$$\text{RMSD} = \left(\sum_s \{f(s, t_i) - f(s, t_j)\}^2 / n_s \right)^{1/2};\tag{7.1}$$

(b) the covariance (from 2.14a):

$$q^a_{ij} = \sum_s f(s, t_i)\, f(s, t_j)/n_s;$$

or

(c) the correlation from (4.1):

$$\rho_{ij} = q^a_{ij} / \sqrt{q^a_{ii} q^a_{jj}}.$$

The RMSD makes a lot of sense, although it may favor states that have small amplitude, where amplitude is defined as

$$\text{AMP}(t_i) = \left(\sum_s \{f(s, t_i)\}^2 / n_s \right)^{1/2}.\tag{7.1a}$$

(Footnote 1 in chapter 2 about weighting applies to (2.14a), (4.1), (7.1), (7.1a).)

The covariance measure, the elements of the, by now, familiar Q^a, is made to regressing each state towards the other so as to minimize their RMSD. We will use here the closely related correlation, but we will not actually modify states by regression. Obviously, a near perfect analogue has low RMSD, and both high covariance and correlation, so the measures should agree in the limit of very good analogues. Many other slight variations in measures of (dis)similarity can be found in Stephenson (1997). In the past circulation types (on the order of 30) were used to quantify similarity (Schuurmans 1973).

In practice we will search for the nearest neighbor in the N-dimensional phase space, and delay answering the question as to whether these nearest neighbors are worthy of the name analogues. The area over which candidate analogues are sought depends on the application. If one wants to compete with global NWP, $f(s, t)$ should include all variables, including boundary conditions and s runs over a global domain. However, in anticipation of the meager results (Lorenz 1969; Ruosteenoja 1988), we scale down expectations considerably, and search for analogues on the area from 20 degrees latitude to the pole, and one variable only, daily $Z500$

at 0Z.[6] This exercise is thus the exact continuation of Chapter 6, where the empirical correlation distribution (ECD) was already studied for the same variable on the same domain.

7.1.2   Search for 500 mb height analogues

As an example of what an analogues search yields, we inspect here the wings of the ECD for naturally occurring analogues on the domain from 20 degrees latitude to the pole, and using just one variable (Z500). Data treatment was already described in Section 6.2, i.e. we first take out a smooth daily Z500 climatology, forming anomalies. First let's focus on just one calendar month, January and one hemisphere, NH. For each date t_i in January there is one other date t_j in January (in a non-matching year!) for which ρ_{ij} is the maximum over all j, denoted max(ρ_i). Averaged over all 31 days and all 37 years we determined that for January (Z500, NH, 0Z data) max(ρ_i) is typically around +0.54. Generally the nearest state is not very near. At 0.5 correlation, the nearest state (and this would be after applying a regression) is at about the distance to the climatology. The record highest max(ρ_i) value in January, 0.71, is between January 25, 1976 and January 10, 1977. While this is better than 0.54 it is only for a single occurrence, and it still is not good enough to be considered very close.

The most dissimilar state with the highest negative correlation is typically −0.51 in January NH. There is a little asymmetry (present in all months and in both hemispheres) implying slightly better analogues than anti-analogues, like +0.54 versus −0.51 in January, NH. This is because high-pressure and low-pressure systems are not exactly each other's opposite (relative to the climatological mean). Indeed a small skew has been noted (White 1980). The record lowest min(ρ_i) value in January is between January 10, 1974 and January 28, 2000 at −0.69.

Figure 7.3 shows the mean value of max(ρ_i) and, with sign reversed, min(ρ_i) along with the absolute highest correlations for the NH in all calendar months. Here we searched only in the same month. The quality of analogues as measured by max (ρ_i) (and anti-analogs as well) is better in winter than in summer. The average correlation one should expect for the best analogue varies from 0.55 in winter to 0.45 in summer, essentially following the inverse of N, compare Figure 6.1. The record best pair in each month has a correlation of about 0.7 in winter (one case > 0.75 in November) to only 0.56 in summer. These "best" lines are more noisy than the average lines because they are based on just one such occurrence in each month—they are the absolute extremes in the ECD. Generally the values for

---

[6] If forecast applications were appropriate here at all, a hemispheric barotropic model is about the most comparable technology.

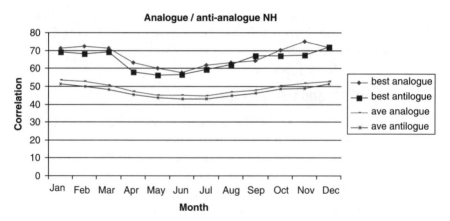

Figure 7.3 The average correlation between a given flow and the nearest neighbor (analogue) and the farthest removed flow (anti-analogue; sign reversed) as a function of month. The record best analogues and anti-analogues are given by the upper lines. The domain is 20°N to the pole, the variable is daily 500 mb height at 0Z, and the years 1968–2004.

analogues are very slightly better than for anti-analogues, both in the mean and the extremes.

Year-round results for the SH, not shown, are similar to those shown in Figure 7.3 for the NH winter. This is because N in the SH is about 30 in any season, see Figure 6.1, i.e. equal to N in the NHs' winter. In seasons other than winter the SH is the better hemisphere for finding (dis)similar patterns because its N is much lower. Smaller N allows wider ECD and therefore slightly higher chance of finding analogues, i.e the average correlation in the SH is between 0.50 and 0.55 year round, and the record best pair by month correlates at better than 0.70 year round, with several record cases ≥ 0.75.

In the above we used data windows of just one calendar month in order to make sure any analogues are during the same time of year, under the same general conditions in terms of solar radiation, SST, etc. However, in the SH winter, both N and the standard deviation of height fields are nearly constant from May to September which is core winter, see Figure 6.1. Using this five-month window we redid the analogue search for the SH. The number of paired comparisons increases with the square of the length of the data set, and 25 times more CPU is used to check all pairs. This extended search netted a handful of additional pairs correlating in excess of 0.75. We also found 20 pairs of anti-analogues that correlate -0.70 or better (but only one beyond -0.75). A search for November to March in the NH yielded only one more case ≥ 0.75 and the two states in this pair are within a month (but not in the same named month).

As a summary all 0Z pairs that correlate better than $+0.75$ for either hemisphere are in Table 7.1. For the 20°–pole extratropical domains there are only 13 such pairs in all of history (1968–2004). The very best case was already shown in Figure 7.1 (where we finessed the search by using 6Z,

12Z, and 18Z data as well). From Table 7.1 we pruned a few cases of >0.75 correlation if those were not the maximum for the two trajectories. For instance, the pair May 10, 1977 and July 31, 1999, which correlate 0.77, is not included in Table 7.1 because this pair follows by one day the May 9, 1977 and July 30, 1999 pair which correlate at 0.80. (One might infer from this that the trajectories were tracking each other for some time and analogues, if and when they occur, make decent forecasts.) As expected the SH yields many more cases above the 0.75 threshold than the NH. We found only two cases better than 0.75 correlation in the NH, the best of which is March 9, 1969 and February 24, 1987, correlating at 0.77.

We also organized the dates of these analogues chronologically in the second entry in Table 7.1. The 13 pairs consist of only 25 distinct dates. Remarkably, September 7, 1996 and August 10, 1974 are both correlated better than 0.75 to September 4, 2000, so we could have a small ensemble (size 2) of forecasts for September 5, 6, etc. 2000. (September 7, 1996 and

Table 7.1. List of 0Z 500 mb height analogue pairs with correlation better than 0.75 on the domain 20 degrees of latitude to the pole. (What is called base and analogue is arbitrary—the reverse also applies.)

	Hemisphere	Date base	Date analogue	Correlation
1	SH	May 9, 1977	July 30, 1999	0.800
2	SH	July 17, 1973	July 12, 2000	0.789
3	SH	May 18, 1968	May 4, 1971	0.787
4	SH	Aug 14, 1981	July 1, 1992	0.779
5	NH	March 9, 1969	Feb 24, 1987	0.774
6	SH	Aug 31, 1989	May 1, 1990	0.772
7	SH	Sept 7, 1996	Sept 4, 2000	0.759
8	SH	Aug 10, 1974	Sept 4, 2000	0.759
9	SH	Jan 19, 1972	Jan 16, 1977	0.759
10	SH	May 23, 1975	May 21, 2001	0.757
11	SH	May 14, 1984	Sept 8, 1996	0.755
12	SH	June 23, 1969	June 30, 1998	0.751
13	NH	Nov 18, 1971	Nov 19, 2004	0.751

Chronologically organized dates, with pair number (1–13) in italics and bold:

1960s	1970s	1980s	1990s	2000s
3 May 18, 1968	*3* May 4, 1971	*4* Aug 14, 1981	*6* May 1, 1990	*2* July 12, 2000
5 March 9, 1969	*13* Nov 18, 1971	*11* May 14, 1984	*4* July 1, 1992	*7,8* Sept 4, 2000
12 June 23, 1969	*9* Jan 19, 1972	*5* Feb 24, 1987	*7* Sept 7, 1996	*10* May 21, 2001
	2 July 17, 1973	*6* Aug 31, 1989	*11* Sept 8, 1996	*13* Nov 19, 2004
	8 Aug 10, 1974		*12* June 30, 1998	
	10 May 23, 1975		*1* July 30, 1999	
	9 Jan 16, 1977			
	1 May 9, 1977			

August 10, 1974 do not correlate very highly to each other (only about 0.6), so the perturbed members in Nature look less like each other than they look like the base, quite the opposite of ensemble members in NWP.) With a little finessing the same could be said about September 7 or 8, 1996, since these 0Z neighbors occur both on the list: May 14, 1984 and September 4, 2000 could be used for an ensemble of forecasts of size 2 for September 12Z, 1996 forward.

While some positive research elements have been noted in the above, the main message is the complete lack of any cases that are truly close, a conclusion reached before by Lorenz (1969), Ruosteenoja (1988), Van den Dool (1994), and many others. In some ways these findings have killed all applications of analogues. In synoptic meteorology, a day-by-day forecast map has been deemed useful as long as the correlation with reality is ≥ 0.6. So to find a pair correlated at 0.80 is better than nothing in terms of pattern similarity and indeed the two maps looks noticeably alike, see Figure 7.1. But in our judgement even a 0.80 correlation is not enough for two states to be called each other's analogue. This is even clearer when we further note that the rms difference for this pair (71 gpm) is still 70% of the climatological standard deviation (100 gpm) in SH winter. In general the distance to the closest neighbor is barely smaller than the standard deviation, i.e. the nearest neighbor is, in general, no closer than the origin (climatological mean). The phase space of N-dimensions dotted with realizations from 1968–2004 is incredibly empty due to undersampling. We have barely begun to observe the atmosphere. Any attempt to study clusters, preferred or non-preferred circulation patterns (on the domain) or analogues will run into serious data limitations unless we can lower N while retaining a meaningful physical system. If we had an analogue pair at the 0.99 correlation level on a large domain, we could veritably test how well NWP, as of today, makes forecasts compared to a perfect model, since they would be starting from similar initial error magnitude.

The situation with respect to finding a natural analogue at a correlation >0.80, > 0.90, etc. on the NH/SH can be described as "possible, but highly improbable", a matter of waiting very long. The problem is not the idea about analogues, the problem is a lack of data in view of the many degrees of freedom in the system.

Table 7.2 shows how the quality of analogues and anti-analogues improves when the search area is reduced progressively. As the area becomes smaller, N decreases as well. For an area as small as 500×1000 km the analogues are near perfect.

7.1.3   How long do we have to wait?

As reasoned in Van den Dool (1994) a three-way relationship can be derived between the size (M years) of an historical data set or library, the distance,

Table 7.2. The average value of max (ρ_i), min (ρ_i) and the value of N for January, 1968–2004, for progressively smaller areas in the NH.

Analogue	Antianalogue	N	Area	Number of gridpoints
53.6	−51.4	29.6	20°N–pole, 0–360°E,	29 × 144 gridpoints
69.7	−68.2	16.3	45°N–50N, 0–360°E	3 × 144 gridpoints
80.2	−78.3	11.4	45°N–50N, 0–180°E	3 × 72 gridpoints
90.7	−88.9	7.0	45°N–50N, 0–90°E	3 × 36 gridpoints
95.3	−94.6	5.3	45°N–50N, 0–45°E	3 × 18 gridpoints
99.1	−99.0	3.7	45°N–50N, 0–10°E	3 × 5 gridpoints

between an arbitrarily picked state of the atmosphere and its nearest neighbor, and the size of the spatial domain, as measured by the effective number of spatial degrees of freedom (N). This heuristic derivation proceeds from the following steps and assumptions.

(i) We assume N equal-variance (carrying sd^2 variance each) processes are going on independently.

(ii) The probability v of two arbitrarily picked states to be within an acceptably small distance ($\epsilon = 15$ gpm would obviously be very good in view of Table 7.1), for a single one of the N processes can be found by integrating a standard normal distribution from $-\epsilon/(\text{sd} \times \sqrt{2})$ to $+\epsilon/(\text{sd} \times \sqrt{2})$. We found $v \approx 0.08$ for $\epsilon = 15$ gpm. An 8% chance is not bad as it practically guarantees an analogue if one has say 100–1000 independent realizations from the past.

(iii) The probability of finding two arbitrarily picked states within tolerance ϵ for all N processes simultaneously is v^N.

(iv) The probability c of finding an analogue in an M year library is $c = 1-(1-v^N)^{20M}$, where the number 20 refers to the number of independent cases in say a two-month window (and M is in units of years).

(v) Demanding $c > 0.95$ leads to

$$M > \ln(0.05)/(20v^N). \tag{7.2}$$

Thus it would take a library of order $M = 10^{30}$ years in an unchanging climate to regularly find two observed flows that match to within current observational error over a large area, such as the Northern or Southern Hemisphere, for just one variable. Van den Dool (1994) gives a table for a range of values of N and ϵ. Obviously, with only 10–100 years of data, the probability of finding natural analogous over a large area (large N) is very very small. For constant M (~40years) and constant ϵ (or v), the annual cycle in N basically explains the annual cycle in the width of the ECD and quality of analogue and anti-analogues as seen in Figure 7.3.

If the N-dimensional phase space has a complicated structure the chance of finding an analogue may depend on where the base case is situated. Nicolis (1998) has shown this to be the case for what is perhaps the most researched dynamical system in history, Lorenz' three variable system (Lorenz 1963). For a system that simple (N is between 2 and 3) one can generate enough data to (a) make analogue forecasting (using data alone, not the equations) a success, and (b) to verify Equation(7.2) or more refined versions.

7.1.4 Application of natural analogues

Because natural analogues are hard to find for large N the applications in geophysics are at best limited. The main trick is to find meaningful physical problems, especially forecast problems, with no more than two to three degrees of freedom. This can be done by the following.

(a) Limit the search area for analogues to a circle of say 1000 km radius (Van den Dool 1989). From the many good matches, see Table 7.2, one can create limited area forecasts which are valid at the center of the circle for short-range ($\leq$12 hours) forecasts. By moving around the circle, a laborious activity, one can make forecasts for large domains, then repeat the process for the next 6 or 12 hour time step.

(b) Use the same 1000 km radius areas, or small enough otherwise, but now downscale or "specify" one field from another (no forward time stepping involved). For instance Kruizinga and Murphy (1983) and O'Lenic and Handel (2004) use limited area height analogues to translate NWP forecasts into surface weather elements. Hamill et al. (2006) use limited area analogues on NWP model precipitation forecasts (of which they have a 22-year homogeneous reforecast data base) to replace historical analogues (to the current forecast) by their matching verification fields (which may be at much higher resolution). Such a practice has also long been in existence in aviation meteorology (Hansen 2000).

Another way of working with just two or three degrees of freedom would be:

(c) Retain only two or three leading empirical orthogonal modes on a domain large in km^2, but small in N. The question is whether this leaves any meaningful forecast problem intact. Given that ENSO ($\sim$ 1 d.o.f.) has such a big influence over at least 20% of the Earth, small natural subspaces $\ll N$, do suggest themselves.

Lowering N by time averaging helps, but not nearly enough. Generating data by very long GCM runs helps (Branstator and Berner 2005), but not nearly enough to make day-by-day short-range forecasts on a global domain possible. (One may also question whether GCM data are a substitute for reality). Researchers in the UK have proposed to run global coupled ocean–atmosphere models for a long time to generate the data from which to select analogue cases for forecasts of ENSO in the future. This could work only if ENSO has effectively only a few d.o.f., which appears to be the case for certain climate variables (Fraedrich 1986). Of course, if it is true that ENSO is contained in just a few d.o.f. one might as well use just the observations, see the discussion in Chapter 10.

Finally, one way out appears to be the construction of an analogue, see next section. Sections 7.3–7.6 are all about applying the constructed analogue.

7.2   Constructed analogues

7.2.1   The idea

Because *natural* analogues are highly unlikely to occur in high degree-of-freedom processes, we may benefit from *constructing an analogue* having greater similarity than the best natural analogue. As described in Van den Dool (1994), the construction is a linear[7] combination of past observed anomaly patterns such that the combination is as close as desired to the initial state (or "base"). We then carry forward in time persisting the weights assigned to each historical case. All one needs is a data set of modest affordable length.

Assume we have a data set $f(s, j, m)$ of, for instance, monthly mean[8] data as a function of space (s), year $(j = 1, \ldots, M)$ and month (m). Given is an initial condition $f^{IC}(s, j_0, m)$, for example, the most recent state (monthly mean map), where j_0 is outside the range $j=1, \ldots, M$, a suitable monthly climatology is removed from the data; henceforth f shall be the anomaly. A (linear) constructed analogue is defined as:

$$f^{CA}(s, j_0, m) = \sum_{j=1}^{M} \alpha_j f(s, j, m) \qquad (7.3)$$

where the coefficients α are to be determined so as to minimize the difference between $f^{CA}(s, j_0, m)$ and $f^{IC}(s, j_0, m)$. The technical solution to this problem is discussed below in section 7.2.2 and involves manipulating the alternative covariance matrix Q^a.

Equation (7.3) is only a diagnostic statement, but since we know the time evolution of the f (we know the next value historically) we can make a forecast by keeping the weights α_j constant in time. More generally we seek a forecast of variable g (which could be f itself) as follows:

$$g^F(s, j_0, m + \tau) = \sum_{j=1}^{M} \alpha_j \, g(s, j, m + \tau). \qquad (7.4)$$

---

[7] We do not rule out that nonlinear combinations are possible, but here we report only on linear combinations.

[8] One can construct analogues for monthly, seasonal or daily data. The procedure is the same. Here we start with monthly.

For $\tau > 0$ we are dealing with a forecast; $\tau = 0$ would be "specification" or down- or up-scaling of g from f (the weights are based on f only!), and $\tau < 0$ would be a backcast. The method is reversible in time. For $g = f$ one can see that the time dependence of f is entirely in the time evolving non-orthogonal basis functions; this is the main trick of the CA forecast procedure and a significant departure from traditional spectral methods in which the basis functions are constant and the time dependence is in the coefficients α_j.

We will later refer to (7.3)–(7.4) in slightly rewritten form, the details depending on whether we use daily, monthly, seasonal or a sequence of seasonal data.

Equation (7.4) can also be written (for $g=f$ and $\tau \neq 0$):

$$f^F(s, j_0, m + \tau) = f^{IC}(s, j_0, m) + \sum_{j=1}^{M} \alpha_j(f(s, j, m + \tau) - f(s, j, m)). \quad (7.4a)$$

In this form the equation looks like a forward time stepping procedure or the discretized version of the basic equation $\partial f / \partial t =$ linear and nonlinear right-hand side terms. Note that on the right-hand side of (7.4a) we make linear combinations of historically observed time tendencies.

Why should Equation (7.4) yield any forecast skill? The only circumstance where one can verify the concept is to imagine we have a natural analogue. That means α_j should be 1 for the natural analogue year and zero for all other years, and (7.4) simply states what we phrased already in Section 7.1 and depicted in Figure 7.2, namely that two states that are close enough to be called each other's analogue will track each other for some time and are each other's forecast. Obviously, no construction is required if there was a natural analogue. But, as argued in the appendix, in the absence of natural analogues a linear combination of observed states gives an exact solution for the time tendencies associated with linear processes. There is, however, an error introduced into the CA forecast by a linear combination of tendencies associated with purely nonlinear components, and so a verification of CA is a statement as to how linear the problem is. Large-scale wave propagation is linear, and once one linearizes with respect to some climatological mean flow the linear part of the advection terms may be larger than the nonlinear terms. This is different for each physical problem.

Is a constructed analogue linear? The definition in (7.4) is a linear combination of nonlinearly evolving states observed in the past. So even in (7.4) itself there is empirical nonlinearity. Moreover, in Section 7.6, we will change the weights during the integration expressed in (7.4a); this will add more to nonlinearity. (We are not reporting on any attempts to add quadratic terms in (7.4); that would allow for more substantial nonlinearity, but the procedure to follow is not yet developed (if possible at all)).

7.2.2 The method of finding the weights α_j

We are first concerned with solving Equation (7.3). The problem is that the solution may not be unique, and the straightforward formulation given below leads to a (nearly) ill-posed problem. Classically we need to minimize U given by:

$$U = \sum_s \{f^{IC}(s, j_0, m) - \sum_{j=1}^{M} \alpha_j f(s, j, m)\}^2.$$

Differentiation with respect to the α_j leads to the equation

$$\mathbf{Q^a \alpha = a} \tag{7.5}$$

This is the exact problem described in Equations (5.1a) and (5.7a). $\mathbf{Q^a}$ is the alternative covariance matrix, $\boldsymbol{\alpha}$ is the vector containing the α_j and the right-hand side is a vector $\mathbf{a}$ containing elements a_j given by

$$a_j = \sum f^{IC}(s, j_0, m) f(s, j, m),$$

where the summation is over the spatial domain. Note that α_j is constant in space; we linearly combine whole maps so as to maintain spatial consistency. Even under circumstances where Equation (7.5) has an exact solution, the resulting α_j could be meaningless for further application, when the weights are too large, and ultrasensitive to a slight change in formulating the problem.

A solution to this sensitivity, suggested by experience, consists of two steps:

(1) Truncate $f^{IC}(s, j_0, m)$ and all $f(s, j, m)$ to about $M/2$ EOFs. Calculate $\mathbf{Q^a}$ and the right-hand side vector $\mathbf{a}$ from the truncated data. This reduces considerably the number of orthogonal directions without lowering the EV (or effective degrees of freedom N) very much.
(2) Enhance the diagonal elements of $\mathbf{Q^a}$ by a small positive amount (like 5% of the mean diagonal elements), while leaving the off-diagonal elements unchanged. This procedure might be described as the controlled use of the noise that was truncated in step (1).

Increasing the diagonal elements of $\mathbf{Q^a}$ is a process called ridging. The purpose of ridge regression is to find a reasonable solution for an underdetermined system (Tikhonov 1977; Draper and Smith 1981). In the version of ridge regression used here the residual U is minimized but subject to minimizing $\Sigma \alpha_j^2$ as well. The latter constraint takes care of unreasonably large and unstable weights. One needs to keep the amount of ridging small. For the examples discussed below the amounts added to the diagonal elements of $\mathbf{Q^a}$ is continued until $\Sigma \alpha_j^2 < 0.5$.

We thus construct the analogue in an EOF truncated space. Strictly speaking we could find an exact solution to Equation 7.3, no ridging, once the fields are truncated to $M/2$ EOFs, with just any $M/2$ years chosen at random. But this procedure would be too sensitive. It is better to use all years, and deal with the underdetermination by ridging.

Instead of using EOFs for truncation the alternative EOT suggest themselves, since we use $\mathbf{Q}^a$. In essence we would rewrite (7.3) as

$$f^{CA}(s, j_0, m) = \sum_{j=1}^{M/2} \gamma_j e(s, j, m) \qquad (7.3a)$$

where the $e(s, j, m)$ are a set of alternative EOTs. One might look upon alternative EOT, orthogonal in space, as the most obvious way of removing collinearity among the $f(s, t)$ so as to make the weights γ_j unique and their calculation easy (just projection). The α_j in (7.5) can be found from γ_j in (7.3a) by a recursive expression. The EOTs are linked to specific moments in time, see Chapter 5, such that the execution of Equation (7.4) is easy.[9] Note that in this variant the base functions are orthogonal initially, but turn non-orthogonal as the forecast proceeds.

One can raise the question about which years to pick, thus facing a near infinity of possibilities to chose from. Here we will use all years. No perfect approach can be claimed here, and the interested reader may invent something better. A large variety of details about ridging is being developed in various fields (Green and Silberman 1994; Chandrasekaran and Schubert 2005), see also the appendix of Chapter 8.

Note that the calculation of $\alpha(t)$ has nothing to do with Δt or future states of f or g, so the forecast method is intuitive, and not based on minimizing some rms error for lead Δt forecasts. There could not possibly be an overfit on the predictand.

7.2.3   Example of the weights

An example of the weights obtained may be illustrative. Table 7.3 shows the weights α_j for global SST (between 45°S and 45°N) in JFM 2000. We have solved (7.3), i.e. found the weights to be assigned to SSTA in JFM in the years 1956–1998 in order to reproduce the SST anomaly field observed in JFM 2000, truncated to 20 EOFs, as a linear combination. For each year we also give the inner product (ip) between the SST field in 2000 and the year in question, i.e. the right-hand side of (7.5). The ip gives the sign of the correlation between the two years. The sum of absolute values of ip is set to 1. Note the following:

---

[9] Because EOTs are not unique it is tempting to use this freedom to think about tailoring EOT for the given base, i.e. find the one state among the $f(s, t)$, say at $t = t_1$, that explains the most of the base, then on to the second tailored EOT, etc. as per the Gram–Schmidt procedure. This yields the set of EOTs that is best suited to explain the base (better even than the EOFs of $f(s, t)$). However, the price to pay is that the historical data set will be truncated more by tailored EOT (than the EOT defined in Chapter 5).

Table 7.3. Weights (×100) assigned to past years (1956–1998) to reproduce the global SST in JFM 2000. The column ip refers to "inner product", a type of weighting that ignores collinearity. See also footnote 12 in section 7.4 on the sequence of four seasons.

yr	ip	α_j	yr	ip	α_j	yr	ip	α_j	yr	ip	α_j	yr	ip	α_j
56	1	18	66	−5	−12	76	2	14	86	3	9	96	3	14
57	0	4	67	−5	−12	77	−3	−2	87	−1	−6	97	2	13
58	−4	−9	68	−2	−16	78	−4	−4	88	0	6	98	−1	3
59	−3	−7	69	−4	− 8	79	−6	−7	89	4	22	99	NA	NA
60	−4	−2	70	−4	− 6	80	−2	−7	90	5	14			
61	2	12	71	2	0	81	−2	−2	91	2	6			
62	−1	−1	72	1	− 2	82	1	−3	92	0	18			
63	−1	2	73	−2	3	83	−1	−5	93	−1	−17			
64	−4	0	74	3	8	84	1	15	94	0	−6			
65	−1	−19	75	1	0	85	5	6	95	0	1			

(i) There is no real large weight, 0.22 being the largest single value, indicating that no year is a natural analogue.

(ii) We allow both positive and negative weights, as the problem is cast in terms of anomalies. A high negative weight would point to an anti-analogue.

(iii) The sum of the weights is unconstrained.

(iv) All years are used, even though years with small weights hardly participate.

(v) The weights are somewhat similar to the ip's (have the same sign usually) but there are exceptions (1992 has zero ip but a high positive weight). These exceptions are caused by the collinearity as expressed by the off-diagonal elements of Q^a, i.e. the fact that years i and j have non-zero correlation.

(vi) Since JFM 2000 was a cold event in the Pacific, the reader may verify that previous cold events generally have positive weight, especially 1989.

(vii) The Pacific SST variability dominates variability globally during ENSO events, so the weights reflect ENSO. Nevertheless, one can also see a trend from mainly negative to mainly positive weights over the 40+ years shown. These trends can be even clearer when the tropical Pacific is quiet.

Below we will discuss three applications to demonstrate how well constructed analogues (CA) work. The first example is specification of monthly mean surface weather from 500 mb streamfunction. In Section 7.4 we describe global SST forecasts; this has been the main application of CA so far. In Section 7.5 we describe how CA works on daily 500 mb height data, and how the dispersion of an isolated source is portrayed by CA, in comparison to EWP. This leads us into the possibility of calculating the fastest growing modes by empirical means in Section 7.6

7.3 Specification or downscaling

Given a 500 mb height map, what is the associated temperature at the same time near the surface? This question comes up when judging the implication of a 500 mb forecast in terms of surface weather. In Figure 7.4, upper left, we have the observed 500 mb streamfunction (ψ)[10] anomaly map for February 1998. Using ψ data for the same domain in the years 1961–1990, and including neighboring months January and March (creating a data set of 90 cases), we first truncated to 50 EOFs (Figure 7.4 upper right), then constructed an analogue, as per Equation (7.3). The result is in the lower left. The EOF truncation is vitally important to find the solution, but as can be seen in the bottom right the error between the upper left and lower left panel is very small. Indeed 50 EOFs reproduce the original field, and

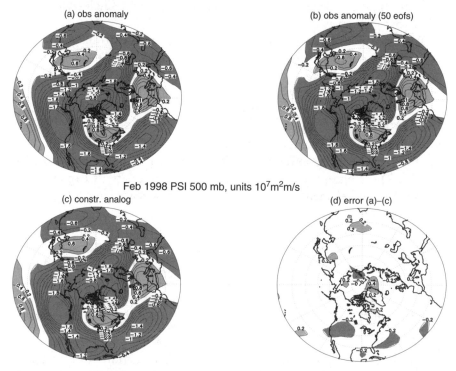

Figure 7.4 (a) The observed anomaly in monthly mean streamfunction (upper left), the same in (b) but truncated to 50 EOFs, the constructed analogue in (c) and the difference of (c) and (a) in (d). Unit is $10^7 m^2/s$, contour interval $0.2 \times 10^7 m^2/s$. Results for February 1998. Domain 20°N–North Pole. Positive values light shading, negative values darker shading. Negative contours are dashed.

---

[10] We present new work and review older work using either a 500 mb height or 500 mb streamfunction. These two fields are closely related in mid-latitude, and nearly equivalent for specification purposes.

the reader will be hard-pressed to see any difference between panel (b) and (c), i.e. within the truncated world the CA is very accurate by design. The error field (a) − (c) is given in panel (d). Making the CA for this application is no problem at all.

With the weights known as per Equation (7.3) we now execute Equation (7.4) for $\tau = 0$, and g is the 850 mb temperature.[11] Note that no temperature fields were used in Equation (7.3). The linear combination of temperature anomalies observed during 1961–1990 is in the lower left of Figure 7.5, and is to be verified against the observed in the upper left of Figure 7.5. One can see that with one set of weights, independent of space, we largely reproduce the near-surface observed temperature anomaly from the streamfunction field aloft for most of the NH. In some small areas the error is considerable, and other factors may determine near surface temperature, but overall the error is 1.0C and the correlation 0.85. We considered many

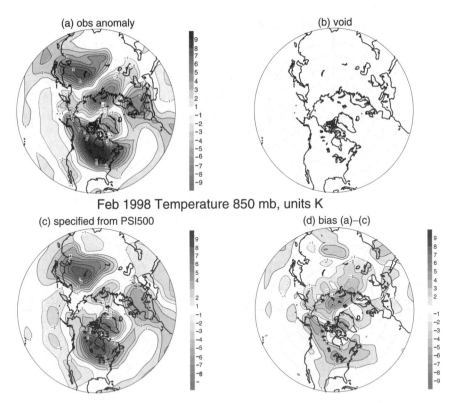

Figure 7.5 (a) (see Plate 10) The observed anomaly in monthly mean 850 mb temperature (upper left), the specified 850 mb temperature by the constructed analogue in (c) and the difference of (c) and (a) in (d). Contour interval 1 K. Results for February 1998. Map (b) is intentionally left void.

---

[11] 850 mb temperature is not literally a surface weather element, but due to difficulties in the Reanalysis/CDAS surface 2-meter temperature, 850 mb is the closest proxy.

cases, and the example shown is typical in performance. In other years we find correlations ranging from 0.65 to 0.90. Specification of precipitation is harder, but still has appreciable skill (not shown).

To be sure, February 1998 is not an arbitrary month since it is during a strong ENSO winter, and the flow pattern in Figure 7.5 may be largely a response to tropical convection. Comparing to the first EOF of ψ in Figure 5.9, one may note high positive projection for 1998. The surface weather in Figure 7.5 resembles regression or composites based on ENSO (Chapter 8).

In Van den Dool (1994) this same procedure was applied to NH 500 mb height, to specify US weather and using CA was shown to work much better than NA. Results over the US by CA appear consistent with far more laborious station-by-station regression equations developed by Klein (1985) for T and Klein and Bloom (1987) for P. Note that CA damps less than regression equations. Still CA damps some, and the error in Figure 7.5(d) (observed minus specified) often has the sign of the observed anomaly.

We thus find that CA works well on problems that are to some good approximation linear. The relationship between streamfunction (or height) and surface weather is fairly linear. As further proof that CA is a very good linear operator indeed we apply CA to the problem of specifying Z500 from ψ 500. This problem presents itself in NWP when at the end of a numerical integration the height field (not a prognostic variable in modern models) has to be derived from ψ. The most complex but still linear relation to obtain Z from ψ is given by the linear balance equation:

$$\nabla \cdot (f\nabla\psi) = \nabla^2(gZ) \qquad (7.6)$$

where f is the Coriolis parameter, and g the acceleration of gravity. ∇ is the horizontal gradient operator. If CA is a good linear operator it should score as high as (Equation) 7.6 on the task of calculating Z from ψ. We tested this for January mean 500 mb data during 1991–2000, 10 cases in all. CA scores better than the linear balance equation in eight out of 10 cases; their average scores are around 0.92 (Equation 7.6) and 0.94 (CA), respectively. Both methods work well, and CA certainly succeeds in being as good as Equation (7.6). This proves we have correctly built a linear operator from data.

The reverse problem, calculating ψ from Z, is more difficult theoretically. But when we used the linear CA to calculate ψ from Z we still found a correlation in excess of 0.90 in all 10 cases.

7.4   Global seasonal SST forecasts

The best prognostic application for CA so far has been the forecast of global SST. This has been done in real time at CPC since about 1993 (Barnston et al. 1994). We use the (near) global SST that has been (is) used as lower boundary condition in the NCEP/NCAR Reanalysis (Kalnay et al. 1996) (its continuation CDAS, Kistler et al. 2001), and form seasonal means. Thus, our data set $f(s, j, m)$ is the seasonal mean SST over the period 1955–present, and m denotes season, $m=1$ for DJF, etc. The most recently observed SST can be approximated by a constructed analogue as per (7.3) as

$$f^{CA}(s, j_0, m) = \sum_{j=1}^{M} \alpha_j f(s, j, m) \qquad (7.3b)$$

where $j=1$ corresponds to 1956, and M is the last year. The weights are determined after truncating the global SST in EOF space. An example of the weights was shown in Table 7.3 when we used data through 1998, and $M=43$. As of this writing we use data through 2003 for the construction, and $M=48$. Given the weights the forecast is given by:

$$f^{F}(s, j_0, m + \tau) = \sum_{j=1}^{M} \alpha_j f(s, j, m + \tau) \qquad (7.4b)$$

which for $m + \tau > 12$ (NDJ) runs into the next year. In the example in Table 7.3 (initial state is JFM 2000), only data through JFM 1998 were used to determine the weights; in Equation (7.4b) the largest forecast lead used is 2 years. If one included 1999 in the construction τ could be only up to one year.

One of the most verified aspects of global SST forecasts is Nino34, an area between 170°W and 120°W and 5°S and 5°N. This index is thought to best describe ENSO (Barnston et al. 1996). Figure 7.6 shows the skill of the CA forecasts for Nino34. The format of this graph is target season vs lead (in months). A zero lead forecast for season 1, DJF, is made at the end of November. The forecasts are automatically cross-validated because the year for which the analogue is constructed has to be left out in Equation (7.3b). For winter target seasons, the anomaly correlation is in excess of 0.9 out to a lead of 2 months, and in excess of 0.6 correlation until a lead of one year. One can clearly see the so-called "spring barrier". For instance, at a lead of 10 months, the correlation for June is < 0.4, but the correlation keeps increasing towards nearly 0.7 at lead 10 through fall and winter until about March, then suddenly drops back. Around the spring barrier the standard deviation is smallest, and most changes of sign of Nino34 take place in NH spring. It is clear that a relatively simple method can have a high level of skill. Earlier verifications (Barnston et al. 1994; Landsea and Knaff 2000; Saha et al. 2006) indicated that CA is among the leading forecast methods, both in real time and on retroactive forecasts. In fact CA stays better than random forecasts until a lead of more than 2 years. A problem in Nino34 forecasting, for all methods, is that there are just a few occasions of strong anomalies in the 50 year record and they make or break the overall verification scores. In between, and this can last for years, none of these forecast methods performs particularly well.

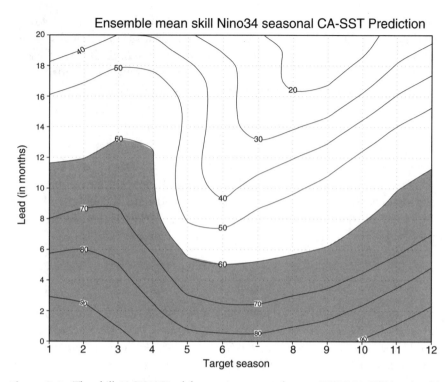

Figure 7.6 The skill (ACX100) of forecasting seasonal mean NINO34 SST by the CA method for the period 1956–2005. The plot has the target season in the horizontal and the lead in the vertical. Example: NINO34 in rolling seasons 2 and 3 (JFM and FMA) are predicted slightly better than 0.7 at lead 8 months. An 8 month lead JFM forecast is made at the end of April of the previous year. A 1-2-1 smoothing was applied in the vertical to reduce noise. Values larger than 0.6 are shaded.

In Equation (7.3b) we construct an analogue to a recent initial condition. As an alternative we have constructed an analogue to a sequence of seasons, i.e. we construct an analogue that is similar to the development of global SST over the whole past year. This does help to boost the skill of the CA method for global SST. The justification for this may be as follows: We are forecasting a single variable from a single variable in a world where many variables are interrelated. An equation for a single variable would analytically be the result of eliminating all the other variables, and the order of the final differential equation may be quite high. This justifies using more that a single initial condition. In effect we use initial conditions for the first, second, third derivative at recent times. To cut down on choices we have used either only a single season (the most recent season) or four non-overlapping seasons (one year).[12] In EOF expenditure terms, these choices

---

[12] Note that Table 7.3 was constructed to the string of MJJ, JAS, OND1999 and JFM2000, each season five EOFs of global SST.

are the same, i.e. we either use a lot of precision in the latest season and none earlier, or we use moderate precision spread out over a whole year. The idea of using analogy over a longer period dates back to the days NA was used (Schuurmans 1973).

The use of four successive seasons also breaks the linearity (or symmetry) of ENSO warm and cold events somewhat. If warm and cold event evolve in different ways, an analogue constructed to a warm event for four successive times will not give high negative weights to a previous cold event.

The verification results shown in Figure 7.6 are actually for an ensemble mean forecast. How is the ensemble made? One can vary the number of EOFs used, here 16, 21 or 26, the participating years as $M+1$ or M, and by spending all EOFs on the latest season's global SST, or spread the EOFs out across four non-overlapping seasons so as to mimic the evolution over a whole year. Any of these perturbations has the effect of changing the weights and creating a somewhat different initial state and a different CA-model. Following these options we obtained 12 members. An example is given in Figure 7.7, which shows the ensemble issued in early July 2005. Half the members use 49(48) years, labeled l for late (e for early) in the third index, and half the members spend the EOF on just the latest season (last 12 months), labeled 1 (12) in the first index. The middle index is 16, 21 or 26 for the number of EOFs used. The forecast by the late members using $M+1$ years stops after one year because the historical evolution following June 2004 was, at the time of forecast release, not known beyond June 2005.

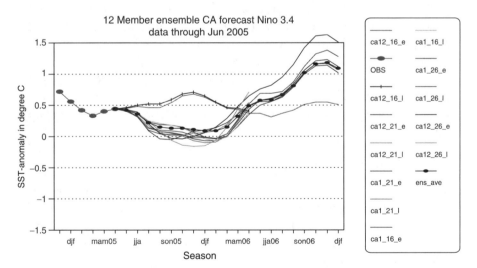

Figure 7.7   An ensemble of 12 CA forecasts for seasonal mean Nino34 SST. The release time is July 2005, data through the end of June were used. Observations (3 month means) for the most recent six overlapping seasons are shown as black dots. The ensemble mean is the black line with closed black circles. The CA ensemble members were created by different EOF truncation etc (see text).

Two members show an increase in Nino34, while 10 members show a tightly clustered decrease. Considerable spread is thus seen to be possible among the members. Traditional statistical methods minimize the rms error in the predictand (here Nino34) and all members would ultimately go to zero anomaly and collapse to zero spread. But even at very long lead the CA members do not agree. Clearly CA[13] has features of error growth, divergence among states which reflects the fact CA has unstable modes. This will be worked out further in Section 7.6.

7.5 Short-range forecasts and dispersion experiments

Short-range forecasting by CA is not a viable practical application at this time because superior methods have been available for decades. The verification presented below is to further understand the strengths and limitations of CA as a method, especially the question as to why it would be useful in the long range and not in the short range. The question of utility for practical application involves comparison to other available tools.

Given is a space–time data set $f(s, t)$; in this case f is the instantaneous daily 500 mb geopotential height taken from NCEP-NCAR Reanalysis, s is a spatial coordinate ($5°$ lat by $10°$ lon grid), and t is time. We form anomalies by subtracting a harmonically smoothed 1979–1995 daily climatology, appropriate for the time of year, produced by Schemm et al. (1997). We consider the domain $20°$N to the North Pole. The data set f is processed as follows: In 1968 we take the fields for January, 1, 3, 5, ..., 23 at 0Z, i.e. twelve fields in one year. Similarly for 1969 through 1992, for a total of 300 fields during 25 years. We now have $f(s, t)$, where $t=1, ..., 300$, representing a great diversity of NH January flows. The time t is a counter for both regular time and annual increments. $f(s, t)$ represents the "library" of historical states used for constructing analogues. A similar data set $f(s, t+\Delta t)$ is used for making the forecast.

7.5.1 Short-range forecasts

We now choose initial conditions every day from January 1–23 in the years 1993, 1994 through 2004, a total of 12 years, having no overlap with the library. Fields are truncated to 100 EOFs (determined from $f(s, t)$), and even though the CA is made in truncated space, the resulting CA field correlates 0.99 with the original. So we have:

$$f^{IC}(s, t_0) \approx f^{CA}(s) \equiv \sum_{t=1}^{300} \alpha(t)f(s,t) \qquad (7.3c)$$

---

[13] The idea of making an ensemble CA was born interactively in a class taught at the University of Maryland with Drs E. Kalnay and M. Cai as participants.

and the forecast is given by:

$$f^{\mathrm{F}}(s, t_0 + \Delta t) = \sum_{t=1}^{300} \alpha(t) f(s, t + \Delta t). \qquad (7.4c)$$

We verified all 276 forecasts (12 years, 23 ICs) based on (7.3c) and (7.4c) for $\Delta t = 1$ day on the domain 20°N to the North Pole. The anomaly correlation is on average only about 0.82, ranging from 0.69 to 0.93 in individual cases. This has to be compared to the mean anomaly correlation of persistence on the same domain, 0.73, and ranging from 0.53 to 0.89 in individual cases.

Even though we have a CA match of 0.99 at the initial time, the correlation for CA drops like a rock to 0.82 in a single day—keep in mind that NWP stays above 0.9 for several days. Perhaps this can be explained away because we did not match the temperature field. But even a barotropic model has above 0.9 correlation at day 1 (Qin and van den Dool 1996). CA and a barotropic model only differ in the formulation of the anomaly vorticity advection by the anomaly wind. Apparently the linear combination of time tendencies associated with nonlinear terms harms the CA forecast in the short-range forecast. Still the total time tendency produced by CA has some skill, otherwise we could not have beaten persistence. It thus follows that even in this short-range forecast problem the linear components of the time tendency are larger than the nonlinear components. Nevertheless, the errors made in the latter put CA behind the competition. Looking back at the high scores for SST forecasts by the CA method months ahead of time (discussed in Section 7.4), one must conclude that SST prediction is a far more linear problem than the short-range weather forecast.

Comparison of CA to NA and EWP can also be drawn. CA is better in forecast skill than NA because the initial match for CA is very good—the day 1 score (0.82) for CA is still much better than initial match for NA (which averages only 0.55 in winter, see Figure 7.3). Even the day 2 score for CA, 0.60 correlation, is better than NA's initial match. The NA may be a perfect method, but its handicap (a bad initial state) makes it unusable; NA cannot even beat persistence averaged over all cases. As a short-range forecast tool the skill of CA is not very different from EWP, compare to Table 3.7 for instance. Saying that the skill is similar does not mean that the forecasts are similar, as we will see in the next section. It may be a partial coincidence that CA and EWP have the same day 1 score. EWP is designed to do a simplified version of wave propagation, an imperfect attempt to calculate the linear part of the time tendency. On the other hand, CA is perfect at the linear part of the time tendency, but adds a presumably bad estimate of the nonlinear term to it.

We also checked the results of CA forecasts for 20°S to the South pole. Results in the SH are entirely consistent with the NH. The gain over persistence is much larger in the SH, roughly as much as it was for EWP in Table 3.7.

7.5.2   CA dispersion experiment

We are now in a position to redo the "rock in the pond" experiments of Chapter 3 to study CA behavior. We will start with the same "round" 500 mb height anomaly disturbance at 45°N used in Chapter 3. But what is different from the EWP dispersion is that the CA dispersion depends on longitude. EWP as designed gives the same result, regardless of the longitudinal position of the initial disturbance. In contrast CA "knows" about the stationary waves in the background field, and implicitly the underlying land–sea distribution. A technical issue is that for EWP we maintained wave amplitude harmonic by harmonic, while CA has an evolving amplitude. We undo this difference by restoring the amplitude of CA to its original value, where amplitude is as defined in Equation (7.1a). Still the spectral distribution may change somewhat.

Figure 7.8 shows the CA dispersion from a source on the date line and 45°N. This has to be compared to Figures 3.1 and 3.2 for EWP1 and 2. The geography in the last two is only for orientation while in Figure 7.8 the geography has real meaning. From the date line and 45°N, the dispersion by CA and EWP are broadly similar, with similar upstream and downstream developments. Mainly zonal dispersion can be seen in CA.[14] The fields after a few days are qualitatively similar, but the details of the far downstream traveling storm tracks show differences in slope, organization, etc. This may be caused by the non-zonal background flow in which the CA dispersion takes place compared to the uniform background flow for EWP.

In Figure 7.9 we compare the two-day forecasts by CA for four positions of an identical rock in the pond at 45°N and the date line, 90°W, Greenwich and 90°E, respectively. The maps are rotated such that the initial source is always at the bottom. We now see that of the four starting longitudes, the dispersion from the date line and Greenwich are the most similar to EWP. The other maps are somewhat similar, but not greatly so. In all cases one can see downstream development of opposite sign, but the intensity depends on longitude. Persistence of the original blob is strongest near Greenwich. Other features are more different. It does matter greatly at which longitude we place the rock, or rather relative to what background flow. Especially the position over Asia gives a different impression.[15] In any

---

[14] This could be in part because we use the domain 20°N–pole for CA, while for EWP2 (spherical harmonics) we use the full globe.

[15] This difference may be slightly overstated because 100 EOFs explain less of the variance of the patch over Asia than elsewhere, i.e. the starting rock in the pond is not 100% identical at all positions once the EOF truncation is applied.

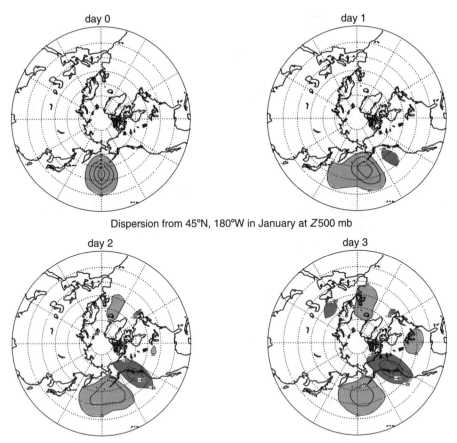

Figure 7.8   Dispersion of the source shown at 45°N and 180°W in upper left, out to 3 days. The dispersion is calculated through an analogue constructed to the initial state. The initial source is identically the same in both shape and magnitude as in Figure 3.1. Contours every 20 gpm. Positive values light shading, negative values darker shading. Negative contours are dashed.

of the four plots there is some slight meridional propagation, the amplitudes being non-zero outside the latitudinal band of the original disturbance, but is not as clear as in Figure 3.2. EWP works with a much more idealized zonally symmetric background flow. A plot similar to Figure 7.9 for the SH has simpler structures, looks more like EWP and shows little dependence on longitude.

7.6   Calculating the fastest growing modes by empirical means

Traditionally instability of atmospheric flow has been gauged by supplying a particular perturbation to a linearized dynamical operator. The operator

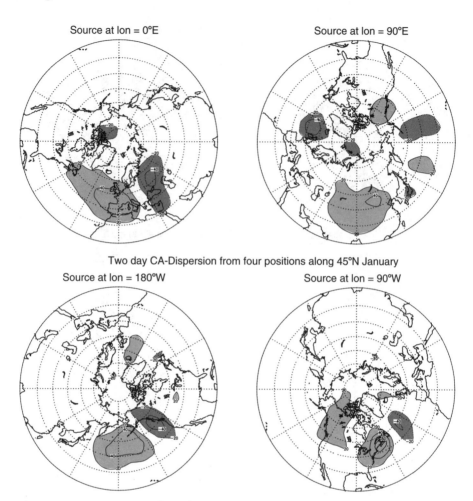

Source at lon = 0°E   Source at lon = 90°E

Two day CA-Dispersion from four positions along 45°N January

Source at lon = 180°W   Source at lon = 90°W

Figure 7.9 Dispersion two days after a sources was released at 45°N and longitudes 0°E, 90°E, 180°W and 90°W, respectively. For ease of comparison the plots have been rotated such that the longitude of release is at the bottom in all four cases. Contours every 20 gpm. Positive values light shading, negative values darker shading. Negative contours are dashed.

is based either on a simplified analytical version of the governing equations, or on a numerical model and a given basic state (usually assumed constant in time). Here we explore the appearance of the fastest growing (or least damped) modes in an operator based on data. The results may apply to any model or data operator in weakly nonlinear conditions. It turns out that familiar low-frequency modes, such as PNA-and NAO-like structures, can be cultured from *daily* data as complex modes with an overall growth rate, a period, two spatial maps, and two associated time series. It is thus suggested that these structures are (almost) unstable modes that grow by drawing energy from the mean flow (a full 3D basic state). As a slight departure from traditional studies we also argue that (1) the time series of the modes, although periodic, do not have to be sine and cosine, and (2) the notion of explained variance in observed data by each mode separately applies under certain restrictions.

As before in Section 7.5, we use the instantaneous daily 500 mb geopotential on the domain 20°N–North Pole. We have $f(s, t)$, where $t=1,\ldots,$ 300 as the historical library. The $f(s,t)$ are also the non-orthogonal basis functions. Formed in similar fashion is a second distinct data set $f(s, t + \Delta t)$, which is the state of field f at a time Δt later; this is where knowledge about time evolution comes in empirically. The time increment Δt is arbitrary, and while we have results for Δt ranging from 6 hours to 5 days, most results shown below are for $\Delta t = 2$ days.

Given an initial condition, $f^{IC}(s, t_0)$ at time t_0 we express $f^{IC}(s, t_0)$, as per Equation (7.3c) as a linear combination of all fields in the historical library, i.e.

$$f^{IC}(s, t_0) \approx f^{CA}(s) \equiv \sum_{t=1}^{300} \alpha(t)\, f(s,t).$$

In order to find $\alpha(t)$ we need to truncate $f(s, t)$ in a reduced degree of freedom space and here we have chosen 50 EOTs (Van den Dool et al. 2000), where 50 EOT functions explain about 87% of the variance in the 0Z unfiltered 1968–1992 data set. (50 EOFs explain 93.5% of the variance.)

Given the initial condition we can make a forecast as in Equation (7.4c) with some skill by

$$f^{F}(s, t_0 + \Delta t) = \sum_{t=1}^{300} \alpha(t) f(s, t + \Delta t).$$

Note that the calculation of $\alpha(t)$ had nothing to do with Δt or future states of f, so the forecast method is not based on minimizing rms error for lead Δt forecasts. This is important in considering growing modes. While all statistical forecast methods, if based on minimizing rms error in the predictand, damp the forecast anomaly amplitude to zero as skill goes to zero, the constructed analogue forecast is unconstrained and produces forecasts with non-zero amplitude indefinitely.

7.6.1 Growing modes

We here address the question as to which structure(s) appear when applying the CA method repeatedly. (The repeated application of an operator is done commonly, see Toth and Kalnay (1997) for example.) The repetition starts by making a new constructed analogue, this time to $f^{F}(s, t_0 + \Delta t)$; this yields a new set $\alpha(t)$ as per (7.3c), which allows us to make a new forecast using (7.4c). Writing F as shorthand for $f^{F}(s, t_0 + \Delta t)$, one can write

$$F_{i+1} = P\{F_i\} \tag{7.7}$$

where forecast number $i+1$ is obtained from forecast number i by applying an operator, which combines steps (7.3c) and (7.4c) into one. Along with F_i we have (t) changing for each i. This process is stable if we renormalize at each iteration, i.e. make $|F_{i+1}| = |F_i|$ (where $| \, |$ is a norm based on Equation 7.1a). This avoids growth to infinity or damping to zero. In the process of expressing F_i at each iteration as a linear combination of $f(s,t)$ we freeze the annual cycle in January, or set the clock back by two days at each iteration (which first goes two days foreward). After many iterations (hundreds or thousands, depending, ...), we save a sequence F_i, where $i = 501$, to 1000 for example, and analyze this synthetic data set by using an EOT analysis. The counter i (iteration) may also be interpreted as time (but now in perpetual season mode), so we analyze a synthetic data set $F(s,t)$. In nearly all cases studied we appear to have converged to a single complex mode $M(s, t)$ which describes $F(s, t)$ in full and can be described as

$$M(s, t) = G[A(s)u(t) + B(s)v(t)] \tag{7.8}$$

where G is the overall growth, $A(s)$ and $B(s)$ are two spatial fields, and $u(t)$ and $v(t)$ are periodic time series multiplying each map. Here t equals the counter i. $A(s)$ and $B(s)$ are determined as the two spatial patterns of the alternative EOTs in the $F(s, t)$ data set, the associated time series $u(t)$ and $v(t)$ are orthogonal (not by construction). Growth G may or may not be exponential ($e^{\sigma t}$) as in Simmons et al. (1983) and Anderson (1991); below G is simply expressed as a percentage amplitude growth per 24 hours. Both the maps and the time series are determined by the process described above. Compared to Simmons et al. (1983), or linear inverse modeling (Penland and Magorian 1993) we have one less constraint, because the time series are not assumed to be sine and cosine. The maps A and B turn out to be spatially orthogonal, while u and v are temporally uncorrelated periodic functions with period T. The instantaneous growth rate defined as $|F_{i+1}|/|F_i|$ before renormalization of $|F_{i+1}|$, while averaging out to the growth rate G, is an arbitrary function of time and is periodic with period $T/2$. The structure that survives the iterations (i.e. comes out first) is either the fastest growing or the least damped mode. The time series look like a deformed sine/cosine pair, i.e. one time series is saw-tooth like, while the other has large residence time at the extremes and fast zero crossings.

7.6.2 Example

An example for $\Delta t = 2$ days may clarify the above. Shown in Figures 7.10(b) are the u and v time series (upper 2 curves) of the first mode in the synthetic data, while the two spatial maps are in Figure 7.10(a). Loosely speaking we go from map A to map B in a quarter period, then to the negative of map A in another quarter, the negative of map B and return to map A after T days, where T may be non-integer. In this case, map A looks

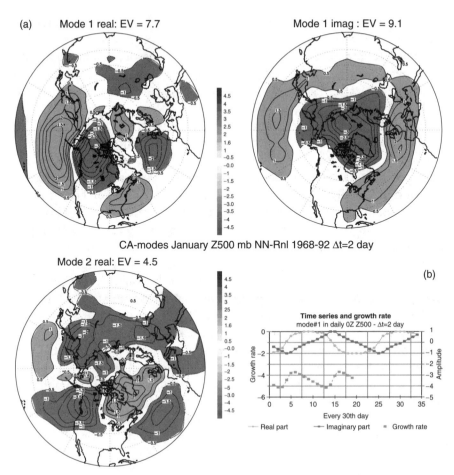

Figure 7.10 (a) The fastest growing modes determined by repeated application of the CA operator for $\Delta t = 2$ days on January 500 mb height data, 20°N–pole. Spatial patterns of the first complex mode are on the top row. The second mode (bottom row) has zero frequency; only a real part exists. Units for the maps are gpm/100 multiplied by the inverse of re-scaling (close to unity usually) applied to the time series in 7.10(b). Contours every 50 gpm. Positive values light shading, negative values darker shading. Negative contours are dashed.
(b) The (u,v) amplitude time series (scaled to ± 1; right-hand scale) and amplitude growth rate (% per day; left-hand scale) of the fastest growing mode, plotted every thirtieth day. The time series and growth rates multiply the patterns shown in the top row of Fig 7.10(a)

somewhat like the PNA, while map B is somewhat like NAO, but we stress that the modes thus produced are not the same as leading EOFs or one-point teleconnection maps. The overall growth rate can be seen in Figure 7.10(b) (lower curve). In the mean this least damped mode decays at 4.5% per day, but the actual damping rate varies between 3.68 and 5.16% per day over $T/2$. The period of this first mode is roughly 630 days, clearly a low-frequency mode, which, rather strikingly, was distilled from daily data arranged in 23-day sequences at two-day intervals from 25 different years. The notion of a period (as in the period of a periodic function) still holds here, although the time series are not single harmonics.

The repeated application of operator P to an arbitrary initial condition, with renormalization at each step, is similar to applying the power method to find the (complex) eigenvector of the (non-symmetric) P. Finding the first mode was described above. We find the second mode by removing from $F(s, t)$ the projection of $F(s, t)$ onto the first mode, and recalculating P, etc. and so on for mode 3, 4, 5 and 6.

Table 7.4 gives pertinent information about the first six modes. The period and the growth rate are classical attributes of normal modes. Here we add explained variance (EV) (in the $f(s, t)$ data), as an interesting and new side issue. One would like to know whether the fastest (least damped) growing modes are a curiosity, or do they really mean something in the world around us. One way of expressing that is to calculate EV, if possible. Luckily by our construction, the real and imaginary parts are orthogonal, and all maps of subsequent modes are orthogonal to the previous maps, such that a unique EV for the real and imaginary part by mode does exist. (The time series of the modes n and m are not orthogonal.) Table 7.4 shows long periods, mildly negative growth rates, and fairly high EV (compare to EOF1–EOF4 in Figure 5.2, which shows 12.2, 8.2, 7.4, and 6.5% respectively). While the order in which modes are selected is determined by growth rate alone, they still, as it happens, order approximately in terms of EV. One might interpret this as a sign that, given random forcing and nonlinearity, the least damped modes have the highest probability to maintain amplitude, and be naturally selected to "explain" variance in the real world. The precise damping rates (5% loss per day) should not be taken literally; for $\Delta t = 6$ hours growth is +6% positive for leading modes.

7.6.3   Discussion of growing modes

We found the resulting modes to be independent of the initial condition (except polarity and phase), i.e. the procedure converges.

Table 7.4. The period (days), the growth rate (% per day) and explained variance (%) in the original (untruncated) data of fastest growing modes 1–6 in January, 0Z 500 mb height, 20°–90°NH, and $\Delta t = 2$ days.

Mode number	Period (Days)	Growth rate (% per day)	Explained variance (%) Real	Imaginary	Cumulative EV (%)
1	630	−4.5	7.7	9.1	16.8
2	∞	−7.8	4.5	–	21.3
3	58	−8.2	4.8	3.9	30.0
4	28	−9.7	2.7	4.5	37.2
5	58	−11.3	5.5	3.6	46.1
6	22	−15.2	2.8	2.2	51.1

Often the period is so long that an exceedingly long synthetic data series would have to be produced to determine whether or not the period is less than infinity. For all intents and purposes we rounded off to zero frequency, or $T = \infty$, if more than 1000 days of integration would be needed. In the case of zero frequency, the oscillation is stuck in one map of fixed polarity plus overall growth or decay. The second mode in Figures 7.10(a) turns out to be zero frequency, and only has a real part. If we had produced data for i=501 to 600 only, we might have concluded that the two maps now combined into one complex mode number 1 (top Figure 7.10a) are two zero-frequency modes. Indeed, when choosing $\Delta t = 1$ or 3 days, the PNA- and NAO-like modes may appear as zero frequency, or be coupled to each other or to yet another pattern, but in all cases the period is very long. With such low damping rates it takes only minimal forcing to cause these modes to persist for a long time.

Sometimes the procedure described above does not appear to converge, i.e. the synthetic data $F(s, t)$ contain more than two EOTs (or more than a single complex mode), no matter how long we iterate. This could either be a failure of the iteration method in a case where the growth rates of two or more modes are very close, or perhaps we need to entertain the thought of generalized modes that consist of more than two maps and two time series. Why not?

The non-sinusoidal character of the periodic time series has been discussed before (Frederiksen and Branstator 2001), the distinguishing feature for this being that the basic state is not an absolute constant (Mak, personal communication), or that the calculation is nonlinear. In Frederiksen and Branstator (2001) the annual cycle was invoked to argue non-constancy of the basic state, but in the empirical approach reported here even the basic state in a perpetual January run would be not be a constant in the sense that energy goes from the basic state to the perturbation and vice versa.

Appendix: Forecasts with CA

1. Intuition

It is intuitively clear that if there was a natural analogue to the base, the forecast would be given by the temporal evolution observed in the analogue year. We extend this here to the intuitive suspicion that once we have the weights of a constructed analogue, the forecast should be based on the same linear combination of the time tendencies observed in the historical years that participate in making the CA. Below we argue that this is exact for linear systems.

2. Formalism

Nature usually satisfies equations like

$$\frac{\partial u}{\partial t} = -u \frac{\partial u}{\partial x} \dots + \text{other linear terms.}$$

Substitution of $u = U + u'$ and subtraction of the maintenance equation for the climatological U yields:

$$\frac{\partial u'}{\partial t} = -U \frac{\partial u'}{\partial x} - u' \frac{\partial U}{\partial x} - u' \frac{\partial u'}{\partial x} \dots + \text{other terms}$$

which can be written in terms of linear and nonlinear operators as:

$$\frac{\partial u'}{\partial t} = \mathcal{L}(u') + \mathcal{NL}(u'). \tag{7.A1}$$

Without loss of generality we could have written down (7.A1) immediately, but we have explicitly emphasized in the lines above (7.A1) that we have linearized the problem as much as we probably can. We shall also assume that (7.A1) represents correctly another source of nonlinearity, namely that arising from forcing. The climatological forcing has been eliminated, and the forcing anomalies can further be split into a part that depends linearly on u' and a nonlinear residual.

Writing the state vector as $u'(t, x, y, z, j)$, where x, y, z are spatial coordinates (dropped from now on), and t is the time of year and j is the year, an analogue is constructed to the base $u'(t,\text{base})$ satisfying to within a small tolerance:

$$u'(t,\text{base}) = \Sigma \, \alpha_j u'(t,j) \tag{7.A2}$$

where the summation is from $j = 1, \dots M$, M being the number of historical cases used.

We now proceed in two ways. First we take the time derivative of (7.A2) without using (7.A1) at all. This yields:

$$\frac{\partial u'(t,\text{base})}{\partial t} = \sum \left(\frac{\partial_j}{\partial t} u'(t,j) + \alpha_j \frac{\partial u'(t,j)}{\partial t} \right).$$

In the CA forecast we neglect any changes of the weights with time, and thus the CA forecast is:

$$\frac{\partial u'(t,\text{base})}{\partial t} \cong \sum_j \left(\alpha_j \frac{\partial u'(t,j)}{\partial t} \right). \tag{7.A3}$$

A forecast can be made. We know the α's from construction at time t, and the time derivative of u' in historical years is known from observations. (The time step is flexible.) Note that the equation (7.A1) is not used so far. A different interpretation of the same procedure, however, results from substituting (7.A2) into (7.A1) but neglecting the nonlinear terms in (7.A1):

$$\frac{\partial u'(t,\text{base})}{\partial t} = \mathcal{L}\left(\sum \alpha_j u'(t,j) \right) = \sum \alpha_j \mathcal{L}(u'(t,j)) = \sum \left(\alpha_j \frac{\partial u'(t,j)}{\partial t} \right). \tag{7.A3a}$$

In (7.A3a), there is no assumption other than linearity. Because (7.A3) and (7.A3a) are identical, it is thus shown that (7.A3) is an exact integration in time for linear systems, and, inter alia, that is constant with time in a linear system. Since the base is an observed field one can easily verify the extent to which α is constant.

In reality the time derivative of u', as observed in years j, does consist of linear and nonlinear tendencies combined inseparably into one. When applying (7.A3) to real data, one obtains, within the accuracy of observation, an exact integration of the linear part of the governing equations (even though the equations are not used explicitly in a numerical scheme). The downside of applying (7.A3) is that a presumably arbitrary linear combination of nonlinear tendencies becomes part of the forecast as well. This introduces error, although it is not clear how much. A different combination of $u'(t, j)$, including requirements for quadratic terms, might yield better results.

8 Methods in Short-Term Climate Prediction

The purpose of this chapter is to list the more common accepted methods used in short-term climate prediction, explain how they are designed, how they are supposed to work, what level of skill can be expected and the references to find more about them. The emphasis is on methodology but aspects of verification and cross-validation will be mentioned as well. Most methods will be accompanied by an example. We will also mention some of the less common methods, but with less detail. We even list some methods that are not used, to delineate which are acceptable and which are not. Sections 8.1–8.6 and 8.8 are easy to read, but Sections 8.7 and 8.9 are more difficult.

It will become clear by the end of the climatology section (8.1), that only the departure from climatology, the so-called anomalies,[1] are considered worthy forecast targets. The climatology itself, including such empirically established facts as "days are warmer than nights", and "winters are colder than summer", is considered too obvious to be a forecast target. This is not to say that a quantitative explanation of the Earth's climate, including daily and annual cycle, is easy. But in professionally honest verification no points are given for forecasting a correct climatology. This chapter is thus about forecasting aspects of the geophysical system that are not so obvious and more difficult. The daily and annual cycle are periodic variations controlled by external forcings such as the solar heating. Implicit in identifying a periodic phenomenon as such is that the forecast of the phenomenon is easy out to infinity. This explains a widespread search for "cycles" in early meteorological research, but very little has been found other than the obvious daily and annual cycles. By removing a climatology that accounts

---

[1] Anomalies, defined as a departure from climatology, have a long history in meteorology. The use of anomalies is very common in modern climate diagnostic studies but it has not always been that way. In the mid-nineteenth century Buys Ballot successfully temporarily championed plotting surface pressure anomalies on weather map so as to make observations taken at different elevations more comparable. The surface pressure anomaly was much later replaced by mean sea-level pressure. The anomaly concept continued mainly in connection with expressing long-range weather forecast. In short-range weather forecasts the use of anomalies is rare, even today.

for daily and annual variations we in effect remove the known easy periodic part of the system.

8.1 Climatology

In the absence of any other information climatology is the best information available. As many travelers can attest, somebody visiting an unfamiliar location 6 months from now is well served by inspecting climatological tables. Climatology is usually defined as a 30-year mean, currently that would be over the 1971–2000 era, while before 2001 it was the 1961–1990 mean. There is nothing magic about 30 years, but it came about (almost 100 years ago) as a long debated compromise in the standards adopted by the WMO, balancing a desire to have much longer averages, on the one hand (which would yield better estimates of the mean, median or "expected value" in a constant climate), and the practicality that most of the homogeneous station records are short, on the other. Note that official WMO climatology is like a discontinuous moving window.

Climatology, as a 30-year mean or otherwise, is often used as the "control forecast" in verification, meaning that any forecast method claiming to be useful should beat climatology in terms of an accuracy attribute A, like rms error (Murphy and Epstein 1989). Skill is usually defined as

$$SS = \frac{A_c - A_m}{A_c - A_p}$$

where the subscripts c, p and m refer to *c*limatology (or *c*ontrol more generally), the *p*erfect forecast and the *m*ethod to be verified respectively. For rms error the attribute A_p is zero and one wants A_m to be smaller than A_c, and the upper limit of skill is 1 (when $A_m = 0$). This is also the exact philosophy underlying the anomaly correlation, see the appendix in chapter 2.[2] Climatology itself then is the base-line forecast with zero skill by construction. In essence one needs to design a verification system that does not give any credit for forecasting climatology thus leaving only departures from climatology a legitimate forecast target. A downside of this strict professional attitude is that if there is no skill by that standard there is a tendency at Weather Services to not issue forecasts at all. After all why publish a forecast without skill? However, many users, especially casual users, need to be reminded that 30-year means are the fall-back

---

[2] A positive anomaly correlation indicates that a properly damped forecast anomaly has lower rmse when verified against the observed anomaly than the zero anomaly climatology forecast. Proper damping in this context means that forecast anomalies are multiplied by the anomaly correlation times the ratio of observed to forecast standard deviations.

"forecast" in that case. Not every user has ready access to this information or understands the situation they are in when Weather Services claim they cannot make forecasts with skill.[3] The 30-year mean is still better than a random guess.

One can obviously use a 30-year data set to learn much more than just the mean. Standard deviation, extremes (records), probability of exceedence for certain thresholds, etc. can be estimated as well. One can define climatology as a pdf, rather than an expected value only. This makes sense also because the profession is moving in the direction of probability forecasts and in that context the control forecast needs to be a pdf as well. Keep in mind that higher order quantities may need many more than 30 years for an accurate estimate, but in a changing climate the expected value loses relevance when evaluated from data extending too far back. This situation is too complicated for anything as simple as WMO standards.

Climatology forecasts are completely local in space. Information from other locations is not used. Creating climatology from neighboring measurements is tricky, especially near orography or other geographical boundaries (land–sea).

Quantitative application of climatology forecasts does require accurate measurements to be taken at the location of interest for many years. In that sense even the most trivial of trivial forecasts does require a sustained effort and does not come for free. And even with all the data at hand, creating climatology from a long and accurately measured time series also has its difficulties. Some authors have recommended smoothing of the data by allowing only a few harmonics, say the annual, semi-annual terms plus maybe one or two more harmonics (Epstein 1981; Trenberth 1984; Schemm et al. (1997)). In doing so the estimate of the standard deviation around the filtered climatology also gets adjusted. These are choices to be made by researchers and practioners.

By necessity, the climatological control forecasts that go along with forecasts issued in real time refer to past climatology (like 1971–2000). For forecasts made retroactively (a model can be run after the fact, say over 1981–2003 for the sake of argument, see Saha et al. 2006), the climatology over 1981–2003 is available. The 1981–2003 climatology is centered in time relative to the retrospective forecast data set, thus creating a climatological control forecast that differs from 1971–2000. This creates some difficulties in comparison of skill estimates and interpretation because the climate may change and the control for verification becomes an issue.

The non-constancy of climatology will be treated in the section on OCN.

---

[3] Additional confusion: Other providers of forecasts may claim skill even if the Weather Service believes there is no skill.

8.2 Persistence

One notch up in complexity and usually skill is a simple method called persistence. In words this means that current conditions are the forecast for a later time. For instance, today's maximum temperature is 27°C, so the forecast for tomorrow is also 27°C. This is essentially a lazy person's forecast with no great expense involved, but because the atmosphere (and especially the ocean) vary slowly, except in special and somewhat rare circumstances, this sort of forecast usually has higher skill than climatology. There is a large collection of related persistence forecasts:

(1) persistence of a previous value, i.e. $f(t)$ is the forecast for $f(t + \triangle t)$);
(2) persistence of the previous anomaly, i.e. $f'(t) = f(t) - \text{clim}(t)$ is a forecast for $f'(t + \triangle t) = f(t + \triangle t) - \text{clim}(t + \triangle t)$;
(3) damped persistence, i.e. $a\, f'(t)$, $0 \le a \le 1$, is a forecast for $f'(t + \triangle t)$.

If the climatology does not change over the $\triangle t$ interval, methods (1) and (2) boil down to the same thing, although the explicit reference to climatology in method (2) may be a helpful reminder to some users. Persistence of the anomaly, method (2), makes clear that the skill of this forecast, when measured by correlation, in the long run equals the auto-correlation:

$$\rho = \Sigma f'(t)\, f'(t + \triangle t) / \sqrt{\Sigma f'(t)\, f'(t) \Sigma f'(t + \triangle t)\, f'(t + \triangle t)}, \qquad (8.1)$$

where summation is over many t (but stratified by time of year). Hence, if the process we are forecasting has a long time-scale, skill will be very high for small $\triangle t$ due to persistence, and may be hard to improve upon for any of the more daring methods. The third version, damped persistence, uses the a priori known skill in formulating the forecast. In order to minimize the rmse of the forecast the factor a in $af'(t)$ should be the autocorrelation times the ratio of the observed to forecast standard deviations.

Clearly the third method is better that the first. This is especially true for large $\triangle t$, say months, when the mean and the standard deviation need to be adjusted, and can be adjusted if measurements have been taken for many years.

Although a persistence forecast does not sound exciting, there have been numerous studies about persistence of both surface weather elements and upper air data (Namias 1952; Dickson 1967; Van den Dool et al. 1986). This is because the degree of persistence varies with location, time of year, and the element studied. A full understanding of such variations is a big challenge. Some oceanographic and atmospheric processes can be closely approximated by a first-order Markov process, so the study of persistence is a study of the whole power spectrum in one number, the autocorrelation at one particular lag (Gilman et al. 1963).

Figure 8.1. shows an example of the annual cycle of persistence of surface air temperature over the USA. The calculations are done with standardized data for many years, placing all 102 climate divisions on an equal footing, then aggregated by summing in space, see Van den Dool et al. (1986) for details. We find that the monthly mean temperature correlates at about + 0.2 from one month to the next (say June to July, called lag 1) and between 0 and +0.15 at lag 2, which would, for example, be forecasting July with May temperature data. Although the correlations are modest, they are positive without exception and the results are undoubtedly statistically significantly non-zero. So in general a month that is above (below) the mean tends to be followed more often than not by positive (negative) anomalies. Thus, there is a tendency for the anomaly to repeat itself. One should note that persistence has a maximum both in late winter and late summer, with lower values in between. The two maxima and minima are typical for the skill of many short-term climate prediction schemes, and in great contrast to NWP, which has a single skill maximum in winter (around the time the degrees of freedom in the atmosphere are lowest) and a single minimum in summer when spatial scales of weather systems are smallest. An attempt to explain the annual cycle in persistence was given in Van den Dool (1983).

Local persistence is much higher along some shorelines. For instance San Diego (USA) and Den Helder (The Netherlands) may have 0.6–0.7 month-to-month correlation at certain times of the year due to the inertia of nearby SST anomalies (Van den Dool and Nap 1985). Schemes that use nearby SST instead of a previous air temperature suggest themselves. One may suspect local effects of this sort near lakes, snow fields, soil moisture, anything at the lower boundary that is capable of extending persistence in anomalies. One should also note that if 2 weeks were the limit of predictability (l.o.p.), as suggested by many fraternal twin NWP experiments, there should not have been any correlation in Figure 8.1 at all for lag 2. The non-zero correlation at lag 2 is thus evidence that the l.o.p. can be

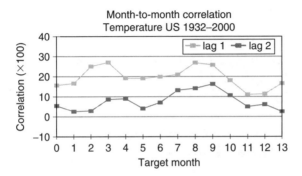

Figure 8.1   Month-to-month autocorrelation ($\times 100$) of monthly mean temperature over the USA at lag 1 and 2 months, as a function of target month.

punctured by the simplest of means. In some cases it is obvious that this is the result of long-lived anomalies in the lower boundary conditions.

One can introduce persistence as a control forecast, and it would be a (much) tougher control than climatology. A strong argument in favor of measuring skill as an improvement over persistence is that the time rate of change is the essence of forecasting, especially if one thinks of the forecast problem in terms of the basic equations with a time derivative on the left-hand side. A zero time derivative is the same as persistence (of type 1), so if we cannot beat persistence the most essential aspect of knowledge about the future is missing.

Just as with climatology, the persistence forecast is purely local, as information of neighboring areas is not used. To apply persistence and especially damped persistence reliable measurements are required, and preferably with more than 30 years of data since calculating the damping factor requires estimates of higher order statistics.

When calculating the damping factor a as a function of $\triangle t$ from data, one may, in some rare cases, discover negative values. This would obviously happen if an oscillatory mode dominates the data and/or the process is clearly not a first-order Markov process. This situation will be covered more extensively when describing regression methods below. But negative a indicates a number of possibilities for obvious extensions, most notably the use of several previous values, e.g. use of both $f(t)$, and $f(t-\triangle t)$ for a forecast valid at $t + \triangle t$.

A source of some debate and confusion is exactly what quantity is, or should be, persisted. Along with the aforementioned methods (1)–(3), should the variable be the very latest value, or the latest weekly mean, the monthly or seasonal mean? Should the length of the time mean of predictor and predictand be the same? And when a (damped) anomaly is persisted, with respect to which climatology is the anomaly taken? Some of the questions raised in the climatology section (8.1) present themselves again. For instance, in Figure 8.1, we calculated persistence in anomalies that are departures from the 1932–2000 mean, but in a real-time forecast during any of these years that would have been impossible since the long-term mean is not known. There are no strict answers, just choices for the researcher and practitioner, choices that need to be justified in some specific context. How about persisting the average over the last 10 years? This will be discussed the next section.

8.3 Optimal climate normals

The definition of climatology as a 30-year mean, adopted by WMO in the early twentieth century, was never fully accepted. This is primarily because the Earth's climate is evidently not constant, especially in temperature. In a slowly varying climate (as opposed to a stationary climate) the average over the last K years may be a better estimate of the upcoming expected value than a longer term mean. OCN stands for optimum climate normals, where the optimum refers to the value of K that minimizes $U(K)$:

$$U(K) = \sum_j \{T_K(j) - T(j+1)\}^2,$$

where $T(j)$ is the temperature in year j, and $T_K(j)$ is defined as

$$T_K(j) \equiv \sum_{j'=j-K+1}^{j} T(j')/K.$$

There is no analytical approach to minimize U. Instead, all values of K are tried. Evaluation of $U(K)$ for each K for temperature over the US (Court 1967/68 ; Huang et al. 1996; Wilks 1996) has revealed that U is a minimum for a value of K on the order of 10 years. There is regional and seasonal variation in K although such variation may not be known accurately enough. The $U(K)$ function is often very flat, such that the optimum K is not that much better than neighboring K values, or has several minima. Moreover, the function U can also be evaluated in terms of absolute error, or in terms of correlation (Lamb and Changnon 1981), leading to similar, but not identical, K. The main point is that K is considerably shorter than 30 years for temperature: $U(K) <$ U(30). Moreover the official 30-year normal is aging until the next update (once every 10 years at best[4]), so that $U(K) <$ U(30) $<$ U(30-fixed period). There are thus two reasons as to why OCN is a better forecast (lower rmse) than the official 30 year normal: (a) $K <$30, and (b) instant update of the K year average.

For certain well-defined processes, one can derive the value of K. For instance in a stationary climate $K = \infty$ (or as close as one can practically be to ∞). For a red noise process $K=1$, using the correlation criterion. For a linear trend (and negligible noise superimposed) $K=1$ also. The empirical value of K is another synopsis of the power spectrum into one number.

The raw forecast anomaly due to the OCN forecast methods is simply the difference of $T_K(j)$ and the official 30-year climatology. This difference, for $K=10$, is evaluated continuously for the USA, and is shown for each rolling season at http://www.cpc.ncep.noaa.gov/products/people/wd51hd/ocn.html. Figure 8.2 shows the OCN temperature maps for the four canonical seasons, as of January 2006. The temperature averaged over the last 10 years is, more often than not, higher than the 1971–2000 normals. This is especially so in the interior south-west USA where all seasons appear to warm and the difference is approaching one standard deviation (of seasonal

---

[4] WMO has had only 1931–1960, 1961–1990 normals. But many countries update every 10 years, so 1971–2000 is in effect as normal in 2006.

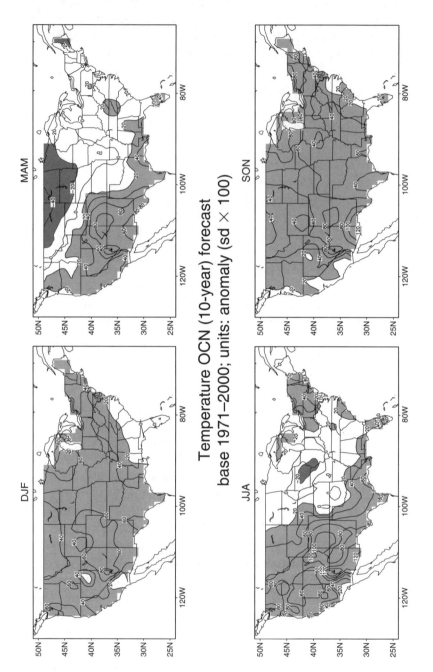

Figure 8.2 The difference between the temperature averaged over the last 10 years and the 30-year climatological mean for the four seasons over the USA. Units are in % of the local standard deviation. Contours every 20. Positive values light shading, negative values (only in MAM and JJA) darker shading. Negative contours are dashed.

mean data around the 71–2000 mean), a huge shift. The only exception to the general warming is an area in the northern plains during spring into summer when temperatures have become lower. Similar maps for precipitation (using $K=15$) are also maintained at the aforementioned website, but the shifts are much less impressive and not nearly as useful for prediction.

OCN has been used internally by certain industries for a very long time to provide a rational basis for calculating prices of such commodities as electricity, heating oil, etc. While OCN is a base-line forecast for climate in the future, OCN had also been used informally in seasonal forecasts at the Climate Analysis Center (now Climate Prediction Center), and at the UK Met Office (Gilchrist 1986). Changes in the Earth's climate, in temperature especially, have become quite evident in the last 20 years. This has made OCN, or other tools that assess "trends", far more important for the seasonal forecast than imagined previously. Since late 1994, OCN has been used in a formal way in the US seasonal forecasts, see Chapter 9. In 1994 we had no idea how important this would be.

OCN is a local method, requiring data only from the site where a prediction is needed. One can look upon OCN either as an amended version of climatology (one that needs continual study and adjustment) or persistence of the average anomaly of the last 10 years. $K=10$ may not stay optimal of course. One may feel OCN is really a forecast for the next 10 years averaged. When verified in that fashion the correlation goes up from 0.3 for single seasons to at least 0.6 for 10-year averaged seasonal anomaly.

Inter-decadal climate variation has been reported in many variables, such as net seasonal Atlantic hurricane activity (Chelliah and Bell 2004) and drought (McGabe et al. 2004). To what extent these OCN-like variations can be described as one phenomenon remains a subject of study. CPC predicts the net seasonal hurricane activity by mainly empirical means and in addition to inter-decadal variations ("hurricane OCN") ENSO plays a role.

8.4 Local regression

The persistence methods, especially the third version, are special cases in the category of local regression, i.e. if $f(t)$ is the predictor and $g(t)$ is the predictand, we seek to determine the coefficients a and c in the expression

$$g^F(t + \Delta t) = a\, f(t) + c \tag{8.2}$$

where $g(t)$ and $f(t)$ are data sets collocated in space. The coefficients a and c can be derived through a regression based on sufficient observations of $g(t + \Delta t)$ and $f(t)$. f and g are time lagged, but the number of time levels is the same for f and g. In the persistence methods discussed above, f and g were the same data set and for $g = f$, Equation (8.2) is autoregression. The intercept c would be zero if f and g are anomalies, and the regression coefficients are calculated over the cases that go into forming the climatology.

Figure 8.3 shows an example. The soil moisture anomaly at the end of the month is the predictor and the predictand is the collocated monthly mean temperature in the next month (lag 1) and 2 months later (lag 2). Soil moisture is not generally observed, but is calculated by a simple physical soil model forced by observed weather elements, integrated forward from 1931 to the present (Huang et al. 1996). The rationale for the exercise is that dry (wet) soil leads to decreased (increased) evaporation[5] and thus increased (decreased) temperature. This effect should only be seen when there is enough incoming solar radiation, since it is the latter that is used either for evaporation or heating the air by the sensible heat flux. Figure 8.3, as was Figure 8.1, is an aggregate result for all of the 102 US "super" climate divisions combined. In general, one can indeed see a negative correlation between soil moisture and subsequent temperature, mainly from April to October, as would be expected. The lag 2 results are similar, but a little weaker. July is more often than not colder than average after May has ended with wetter than average soil. The effect of soil moisture may explain in part why temperature itself is persistent in summer (Huang and Van den Dool 1993).

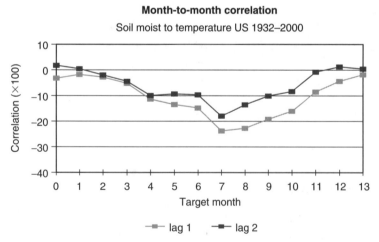

Figure 8.3 The local correlation ($\times 100$) between antecedent soil moisture and subsequent temperature over the USA. Results are aggregated over 102 super climate divisions.

---

[5] By evaporation we mean the combined evaporation from all sources, i.e. from bare soil, bodies of water, water on the canopy as well as transpiration by green vegetation during daylight hours.

While the correlation is negative in Figure 8.3, the skill of a forecasting scheme based on soil moisture is the absolute value of the shown correlation as long as results hold up perfectly on independent data. Because we used a lot of data (70 years, 102 locations) that may be very nearly the case here. The degradation of a full sample correlation (ρ_{fs}) in a CV-1-out test, is given by (Barnston and Van den Dool 1993)

$$\rho_{cv} = \frac{L-1}{L} \rho_{fs} - \frac{1}{\rho_{fs}(L-1)} \qquad (8.3)$$

where ρ_{cv} is the correlation after cross-validation is done, and sample size $L = N \times M$, the product of N, the number of degrees of freedom in space (which is 5–15 for temperature in the USA as evaluated from climate divisions (Huang et al. 1996) and M (the number of years if year-to-year autocorrelation is small). Shown in Figures 8.1 and 8.3 are full sample correlation (ρ_{fs}) and ρ_{cv} as per Equation 8.3 would be the correlation of forecasts against observations to be expected on independent data. The two terms in (8.3) combine shrinkage of the correlation as well as degeneracy of the CV-1-out procedure. For local correlation, L is at most 70, and a full sample correlation of 0.30 for example should not be expected to yield more than 0.24 when applied as a regression forecast. The spatially aggregated correlations are expected to hold up better on independent data because L is much larger.

Aggregation in space of local results in Figures 8.1 and 8.3 makes sense only because we expect somewhat similar effects everywhere, and the aggregation greatly reduces the noise. Aggregation would make no sense at all if soil moisture correlated positively (negatively) with subsequent temperature in one half (the other half) of the domain. Figure 8.4 suggests some structure in the time-lagged soil moisture–temperature correlation, but the sign of the lagged correlation is negative nearly everywhere. We have chosen spring in Figure 8.4 because this is a very interesting season. The state of soil moisture at the end of February has only a small impact in March (upper right), but suddenly grows in importance in April (lower left) in the south–east USA. This may be related to the greening of plants as the season advances; after all, most of the evaporation comes through the vegetation. Note also that the simultaneous correlation (upper left) is positive in a few areas. This happens in winter in areas where precipitation is associated with warm air advection. But correlations are negative nearly everywhere at positive lag.

No one stops us from considering a local regression involving yet another data set, or several others:

$$g^{F}(t + \Delta t) = a\, f(t) + b\, h(t) + c \qquad (8.2a)$$

where g, f and h could be collocated temperature, soil moisture and snow depth, for instance. The problem with having to estimate three or more coefficients (as a function of space, month and Δt) is having enough

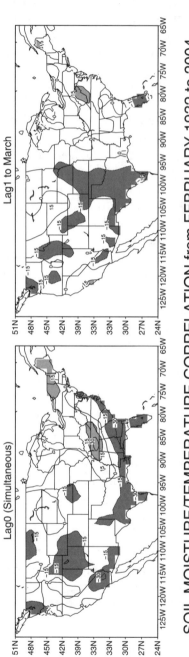

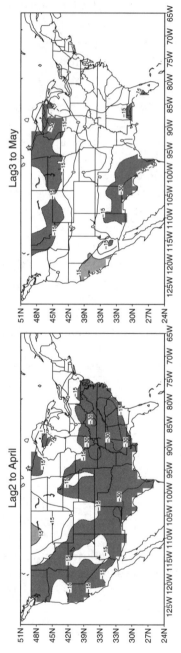

Figure 8.4 The correlation (×100) between soil moisture at the end of February and the collocated temperature in February (lag 0) through May (lag 3). Correlations in excess of ± 0.15 are shaded. Positive values light shading, negative values darker shading. Negative contours are dashed.

observations to make reliable estimates. In short-term climate prediction we rarely have a lot of data so prudence is very important. It helps to have a physical basis for picking a certain predictor. Trying out every predictor under the sun is a certain disaster.

A special case of (8.2a) would be:

$$f^F(t + \Delta t) = a\, f(t) + b\, f(t - n\Delta t) + c \qquad (8.2b)$$

where f is forecast from several previous values of itself. It does happen in some rare instances that a, when determined from regression, goes negative for certain values of Δt. As a curiosity Namias (1952) and Dickson (1967) would report "anti-persistence", i.e. a tendency for the anomaly to change sign at the chosen lag. If reliable, this would indicate a sufficient departure from a first-order autoregressive to admit one or more previous values. For instance if $f(t)$ were a sine wave the best prediction would be to use the latest value and another value one quarter of the period earlier. A great example, perhaps the only one, is the Quasi-Biennial Oscillation.

Figure 8.5 shows the time series of the zonal mean zonal wind at 30 mb for the period 1948–present. The somewhat cyclical variation, known as the QBO in the stratosphere, invites a scheme like (8.2b) for its prediction, where $n \times \Delta t$ should be about one quarter of the period.[6] Prediction of the QBO has been deemed important for seasonal hurricane activity forecasts (Gray et al. 1994). The forecasts of the QBO can be made using the naked eye, i.e. it is nearly obvious for a few seasons out. However, the period ("quasi-biennial") is not exactly constant and the amplitude varies as well, so forecast schemes of the QBO are not as successful (or trivial) as a forecast of the ever-repeating annual cycle or the tides in the ocean. The QBO, in spite of its near regularity, is not known to be linked to external forcing, but is caused by dynamics internal to the atmosphere (Holton and Lindzen 1972; Plumb 1977). The phenomenon was discovered in the late-1950s (Reed et al. 1961). The QBO is by far the most predictable unforced signal in the atmosphere, very much beyond the two-week limit of predictability, but its impact on surface weather and climate remains largely unclear. Moreover, there may be weaker biennial signals of a different origin near the surface (Barry and Carleton 2001).

Note that at the beginning of the graph in Figure 8.5 the QBO appears absent. In the early years, 1948 to mid-1950s, there were no observations that high up in the stratosphere to be assimilated in the Reanalysis; in reality there probably was a QBO even though Figure 8.5 does not show one. The Reanalysis has yielded a beautiful and accessible data set, but one can never take data for granted.

---

[6] An alternative is to use the zonal wind simultaneously at various levels in the vertical because the QBO signal is known to propagate downward.

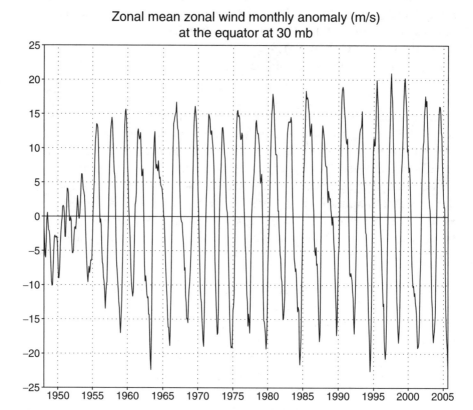

Figure 8.5 The zonal mean zonal wind at 30 mb at the equator from 1948 to September 2005. The time mean is removed. All data is monthly mean. Source is the global NCAR/NCEP Reanalysis

The QBO may have high negative correlation at a lag of about 13 months, but closer to the ground anti-persistence in wind, temperature or pressure is always small and may be sampling error. If anti-persistence were common, persistence would not be a recommended control forecast, and would not be an improvement in skill relative to climatology. There are only a few locations near the storm track where the passage of cyclones is regular enough to suggest a modest negative autocorrelation in wind and pressure after a few days. Van den Dool and Livezey (1984) report on the spatial distribution of persistence of monthly 700 mb height anomalies. The lag 1 autocorrelation is generally positive, especially at low latitudes, but there is a puzzling case of month-to-month anti-persistence in spring in the mid-latitude Atlantic troposphere. Synoptic experience also suggests major changes in the Atlantic area can take place in that season. Black et al. (2005) have tried to explain this by up-and-down propagation of NAO-like signals into the stratosphere and back, and linking spring anti-persistence near the surface to the process of "final warming" of the stratosphere.

One could give the procedure of local regression an analytical twist by assuming a "model". For instance in Equation (8.2), when written as $f^F(t + \Delta t) = a \times f(t) + c$, the value of $a(\Delta t)$ could be either calculated for each Δt, or we calculate $a(\Delta t)$ for only one specific value of Δt, but generalize the resulting a by assuming how $a(n \times \Delta t)$ relates to $a(\Delta t)$. For instance, in a first-order Markov chain the autocorrelation drops off theoretically as $\rho(n \times \Delta t) = \rho^n(\Delta t)$.[7] One Δt calculation suffices, or one could smooth the results a bit from an evaluation of coefficient a for several values of the time lag. This semi-analytical approach will be important later on in the discussion of LIM.

8.5   Non-local regression and ENSO

It is methodologically obvious how to pursue non-local regression. In simplest form we have

$$g^F(s_1, t + \Delta t) = a \times f(s_2, t) + c \tag{8.4}$$

where we seek to forecast g at location s_1 from the predictor f at location s_2. Position s_1 may not be arbitrary, since it is the location for which a forecast is desired, but the choice of s_2 may be more difficult. There has to be a good justification and/or thorough statistical testing. Perhaps the best known example of a relationship that fits Equation (8.4) is given in Figure 8.6 (as well as on the cover of the book) which shows the correlation between Nino34 (or Darwin pressure) in SON and the seasonal mean temperature and precipitation over the USA in the subsequent JFM period. The calculation is based on data during 1931–2004. The ENSO related teleconnection is thought to explain this pattern. While teleconnections are near-simultaneous, anomalies in tropical indices like Nino34 are long lived, and the lagged correlation displayed in Figure 8.6 is thus about forecasting a teleconnection several months ahead of time. In Figure 4.2 we showed a one-point teleconnection pattern for 500 mb height in JFM which may accompany the surface weather elements shown in Figure 8.6. (Figures 7.4 and 7.5 give a specific example for February 1998, which conforms well to the canonical pattern during ENSO.) There is no such 500 mb pattern teleconnection during SON itself. But communication with the tropics opens up when the westerlies advance farther to the south with season (Opsteegh and Van den Dool 1980) and the teleconnection emerges.

What is especially noteworthy about Figure 8.6 is that the ENSO impact on precipitation is about as large (or larger) as the impact on temperature. It

---

[7] Lorenz (1973) noted that autocorrelation showing positive departures from a first-order Markov process at longer lags is basic empirical evidence of longer range predictability.

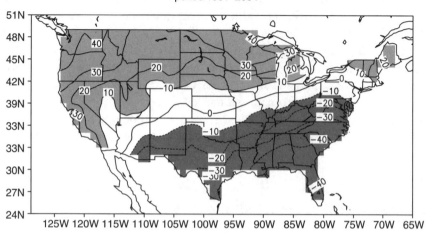

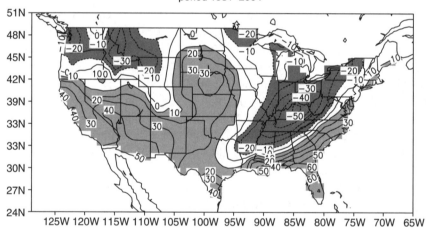

Figure 8.6 The correlation (×100) between the Nino34 SST index in fall (SON) and the temperature (top) and precipitation (bottom) in the following JFM in the US. Correlations in excess of ±0.1 are shaded. Contours every 0.1. Positive values light shading, negative values darker shading. Negative contours are dashed. Color version of this figure is on the cover of the book.

is highly unusual for mid-latitude forecasts of precipitation to have skill anywhere near the skill level of temperature. (In fact we did not show the companion results in Figures 8.1 and 8.3 for precipitation because the correlations are indistinguishable from zero.) Here is the one exception. During ENSO warm events the southern USA, and especially the south-east and interior south-west are likely to be wet. In south Florida the correlation is better than 0.60. Note also the gradient in the correlation between Kentucky and Florida, indicative of a change in storm track across the south-east USA. The temperature signal (upper panel in Figure 8.6) is for positive anomalies across the north and negative anomalies in the south-east when Nino34 is positive. Maps based on Nino34 and SLP at Darwin as predictor during SON look very similar. Darwin pressure is one of two pieces of input to the SOI, and the SOI is closely related to the ocean's SST. That the measurement of pressure at a single location in Australia (or SST in a small area) would have such far reaching prognostic consequences is among the more exciting aspects of our profession. It also seems odd that we can make such statements based on very simple means, given that forecasting the weather a few days out requires detailed input initial data on a global domain. How can short-term climate prediction be so simple?

The impact of ENSO on climate far away was not known very well before about 1980, in part because global data sets had not been available in real time before 1980. Douglas and Engelhart (1981) reported a high correlation between autumn precipitation on islands in the Pacific and winter rainfall over Florida, much the same as shown in Figure 8.6. Madden and Van Loon (1981) reported on *simultaneous* global effects of a measure of the Southern Oscillation and pressure and temperature elsewhere, wherever there was data. The strong 1982–83 event, along with the availability of global data (satellites as well), prompted interest that has remained high ever since. The ENSO teleconnection research is often about simultaneous relationships, the assumption being that because of high persistence in the tropics any simultaneous relationship carries over at lag.

In spite of the long-distance ENSO impact over land, even more so in Australia, Indonesia and South America than over North America, one should not lose sight that the correlations (over the USA) are still modest with absolute value exceeding 0.4 in only limited areas and mainly in winter. Most of the area shown in Figure 8.6 has correlations less than ± 0.2, which means that most of the USA does not have a significant linear correlation with ENSO. Mason and Goddard (2001) come to the same sobering conclusion for the entire globe. Linear correlation has obvious drawbacks. More on ENSO will be discussed in the section on composites.

Except for links to ENSO, there may not be that many non-local simple prognostic relations that fit Equation (8.4). However, if we relax the definition of spatial position and allow $g(s_1, t)$ and $f(s_2, t)$ to be the sine and cosine components of moving waves, EWP (which could be named Lagrangian persistence) may be looked upon as the complex version of Equation (8.4), i.e.

$$g^{F}(s_1, t + \Delta t) = a\, f(s_2, t) + b\, g(s_1, t)$$

$$f^{F}(s_1, t + \Delta t) = c\, f(s_2, t) + d\, g(s_1, t)$$

which, as shown in Chapter 3, when executed wave by wave, leads to the expressions for empirical wave propagation. For unfiltered daily data EWP has some skill, but no prognostically useful phase propagation is known to exist for filtered monthly or seasonal mean data.

For those seeking more references on methods in non-local regression, there is a body of literature on model output statistics beginning with Glahn and Lowry (1972).

8.6 Composites

A composite is an operation on a subset of the available data, which is conditioned to satisfy some criterion. For example, one can calculate the mean winter temperature in Washington DC for those years when DJF Nino34 was at least 0.5°C above the mean. In the same way, one can make probability statements: during years when Nino34 SST was at least 0.5°C below the mean, the precipitation at some locale in the southern USA was 8 out of 10 times in the below-normal tercile. Such composites should be broadly consistent with Figure 8.6 but allow for (a) an evaluation of asymmetry (nonlinearity) in the relationship between Nino34 and remote weather and climate, and (b) an understandable user friendly forecast. Is the impact of a La Nina the exact opposite of an El Nino as assumed in linear correlation? If a cold event is expected for next winter, the statement that "8 out of 10 previous cold events were particularly dry" is clear to even the person in the street. Of course one can make composites based on any criterion. The reason ENSO is used here as an example is because Nino34 is (i) quite predictable months ahead of time at most times of the year, and (ii) has discernable influence. Composites as a technique are widely used for diagnostic research. One may study the mean weather conditions given that the PNA or NAO are simultaneously far above normal. In many places in Europe and North America, the PNA and NAO have a larger influence than ENSO but because prediction of NAO and PNA is far less successful, an NAO composite is mainly a diagnostic tool. In Chapter 4, we used compositing to study asymmetry in the relationship between base points near Greenland and Europe. Research into the skew, asymmetry and or nonlinearity of ENSO includes work by Burgers and Stephenson (1999), Montroy et al. (1998), Hoerling et al. (1997), Lin and Derome (2004), and Wu et al. (2005).

ENSO composites have been a standard tool in long-range forecasting since Ropelewski and Halpert (1986, 1987, 1989) and Halpert and Rope-lewski (1992) made their famous display of world-wide short-term climate anomalies related to the Southern Oscillation. However, the details of the compositing method have changed since then. There does not appear to be a settled method. At the CDC website http://www.cdc.noaa.gov/index.html one can make composites for the USA "on demand" in an interactive database and graphics system. On the CPC website one can find a fixed set of US composites used by its forecasters at http://www.cpc.ncep.noaa.gov/products/precip/CWlink/ENSO/total.html. However, even here the number of options is large, due to trend adjustment considerations (adding OCN into the composite; see Higgins et al. 2004). The composites used for the famous 97/98 ENSO winter can be found in Barnston et al. (1999). More on composites can be found in Wolter et al. (1999) and Cayan et al. (1999).

Dangers with composites are obvious. Since they are based on fewer cases (than a linear regression using all data), the result is noisier and may be overinterpreted. Sampling variability may be confused with true nonlinearity. Also, minute changes in the threshold may change the number of cases (and the results) significantly. The criterion for ENSO is debatable, but in recent years NOAA has adopted a 0.5°C Nino34 anomaly criterion in three month running mean Nino34 SST for its real-time guidance.[8] The advantage and simplicity of composites is quickly lost when more than one factor or other considerations enter the composite.

8.7   Regression on the pattern level

Most empirical methods in short-term climate prediction are nowadays based on multiple linear regression "on the pattern level". A primitive example is as follows. Suppose we have two data sets, $f(s, t)$ called the predictor, and $g(s, t)$ called the predictand. One can perform two stand-alone EOF analyses of f and g, and then do the regression between the time series of the leading modes in the predictand and predictor data sets. Klein and Walsh (1984) made an in-depth comparison of regression between EOF mode time series on the one hand and regression between the original data at grid points on the other; this was in the context of "specification" (as discussed for instance in Section 7.3). Using modes is efficient, and cuts down on endless choices, but it may not always help the skill.

For a more general approach we first discuss the time-lagged covariance matrix.

8.7.1   The time-lagged covariance matrix

When we have two data sets, $f(s, t)$ called the predictor, and $g(s, t)$ called the predictand, one can define the elements of the time-lagged covariance matrix $\mathbf{C}_{fg}$ as

---

[8] For historical research purposes warm or cold events should have at least five successive rolling seasons in excess of the 0.5 criterion.

$$c_{ij} = \sum_{t=1}^{n_t} f(s_i,\, t)\, g(s_j,\, t+\tau)/n_t \qquad (8.5)$$

where n_t is the number of time levels, a time mean of f and g was removed, and τ is the time lag. C, non-square and asymmetric in general, thus contains the covariance between the predictor at any place in its domain, and the predictand anywhere in its domain, local as well as non-local. (From the time lag in g our intention is clear: to predict g from f. However, some analyses below (CCA, SVD) do not go beyond establishing associations between f and g, leaving in the middle who predicts whom. Most texts on SVD and CCA thus do not show a time lag.)

Associated with c_{ij} there is also a correlation

$$\rho_{ij} = n_t\, c_{ij} \Big/ \sqrt{\Sigma f^2(s_i,t)\, \Sigma g^2(s_j,t+\tau)}. \qquad (8.5a)$$

If g and f were the same data set, and the time lag is zero, C would be the square Q, as per Equation (2.14). Along with C_{fg} we also need Q_f and Q_g below; $Q_f = C_{ff}(\tau = 0)$. Given how Q was manipulated to calculate EOF (presented in Chapter 5 as "self prediction") one may surmise that C can be used to relate patterns and time series in the predictor field to patterns and time series in the predictand field. Indeed, C, in its various renditions depending on prefiltering, truncation, orthogonality constraints, organization of input data sets, etc. is among the most studied in short-term climate prediction. Instead of the role played by the notion explained variance (EV) in EOFs, the target of calculating coupled patterns/time series is often in "explaining" the covariance of f and g. Because covariance can be negative, the target is often taken to be "squared covariance" (SC), i.e. the fraction of Σc^2_{ij}, where summation is over all i and j, that can be explained by 1, 2 or m coupled "modes".

Without any truncation or constraint C is set up to create any imaginable regression between f and g, so as to minimize the rmse of the prediction of g, on the dependent data that is used to compute C. Here lies a very significant problem. With so many predictors $f(s, t)$, it is hard to avoid overfitting.[9] C contains the correlation of everything with everything. The overfit is combated by severe truncation at the pattern level. This reduces the subjective nature of choosing predictors.

Somewhere in C also lie the methods we already discussed before, like local persistence and local and non-local regression. The reason to present these simpler methods separately and upfront is twofold. First we may easily lose local effects when applying truncation at the pattern level, i.e. the very high persistence in temperature in San Diego, California would not make it into a pattern method until hundreds of modes are

---

[9] Standard texts on regression should be consulted to find methods of exploratory regression that can avoid overfit in most cases.

admitted. Secondly, **C** is calculated without any physical intuition. The local effects approach can more easily be defended on physical grounds.

8.7.2   CCA, SVD and EOT2

In Chapter 5 we presented EOFs of the data set $f(s, t)$ as:

$$f(s, t) = \sum_{m=1}^{M_f} \alpha_m(t)\, e_m(s) \tag{8.6}$$

where both the time series and spatial patterns are orthogonal. Equation (8.6) still gives a complete representation of f as long as either the time series or the spatial patterns are orthogonal, and M_f is large enough. Likewise we have for the predictand:

$$g(s, t + \tau) = \sum_{m=1}^{M_g} \beta_m(t + \tau) d_m(s). \tag{8.6a}$$

Coupling the modes among the two data sets f and g, which have the same number of time levels but possibly different spatial domains (also $M_f \neq M_g$), will be discussed below in terms of the properties of $\alpha_m(t)$ and $\beta_m(t + \tau)$ and $d_m(s)$ and $e_m(s)$, respectively. In any of the methods below orthogonality is maintained in either time or space (not both), so the coupled modes allow projection of future data and/or partial rebuilding of f and g themselves with a set of modes, and the notion explained variance (not optimal obviously) within each data set still applies.

The plain distinguishing feature of Canonical Correlation Analysis (CCA) is that the correlation of $\alpha_m(t)$ and $\beta_m(t + \tau)$, denoted cor(m), is maximized; the modes are ordered such that cor(m)>cor($m+1$) for all m. Within each data set we have for CCA

$$\sum_t \alpha_k(t)\, \alpha_m(t) = 0 \qquad \text{for } k \neq m \tag{CCA-1}$$

$$\sum_t \beta_k(t + \tau)\, \beta_m(t + \tau) = 0 \quad \text{for } k \neq m \tag{CCA-2}$$

i.e. orthogonal time series, and across the data sets:

$$\sum_t \alpha_k(t)\, \beta_m(t + \tau) = 0 \qquad \text{for } k \neq m \tag{CCA-3}$$

$$\sum_t \alpha_k(t)\, \beta_m(t + \tau) = \text{cor}(m) \quad \text{for } k = m \tag{CCA-3a}$$

where summation is over time. The cor(m) can be found as the square root of the eigenvalues of the matrix $\mathbf{M} = \mathbf{Q}_f^{-1}\mathbf{C}_{fg}\mathbf{Q}_g^{-1}\mathbf{C}_{fg}^T$ (or from $\mathbf{Q}_g^{-1}\mathbf{C}_{fg}^T\mathbf{Q}_f^{-1}\mathbf{C}_{fg}$). Note that CCA maps are not orthogonal.

On the other hand, in a method often called singular value decomposition (SVD) the explained SC is maximized. For SVD[10] we have within each data set:

$$\sum_s e_k(s)\, e_m(s) = 0 \qquad \text{for } k \neq m \qquad \text{(SVD-1)}$$

$$\sum_s d_k(s)\, d_m(s) = 0 \quad \text{for } k \neq m \qquad \text{(SVD-2)}$$

i.e. orthogonal maps, and across the data sets:

$$\Sigma \alpha_k(t)\, \beta_m(t+\tau) \quad = 0 \qquad \text{for } k \neq m \qquad \text{(SVD-3)}$$

$$\Sigma \alpha_k(t)\, \beta_m(t+\tau) \quad = \sigma\,(m) \quad \text{for } k = m \qquad \text{(SVD-3a)}$$

where $\sigma\,(m)$ is the mth singular value of C_{fg}. The SC explained by mode m is $\sigma^2(m)$.

Notice the (dis)similarities of SVD and CCA. CCA has orthogonal time series, SVD orthogonal maps. Properties (CCA-1) and (CCA-2) vs (SVD-1) and (SVD-2) appear to be a matter of space–time reversal, but this cannot be stated for the third property. The roles of cor(m) and $\sigma(m)$ appear similar. The notion "SC explained" is sometimes also used for CCA, but does not relate trivially to cor(m). Theoretically it is possible that the first CCA mode describes a perfectly coupled f-g process of infinitesimal amplitude (high cor, low SC).

CCA and SVD are methods to find coupled modes, but they are not quite forecast methods. A regression between the $\alpha_m(t)$ and $\beta_m(t+\tau)$ is needed to forecast $\beta_m(t+\tau)$ given $\alpha_m(t)$.

An easy way of explaining both the idea and the actual application of methods like CCA and SVD to a forecast situation may be to use "EOT2"; we used EOT in Chapters 4 and 5, but extend it here to two data sets. Specifically, we seek the position s_1 in space so that the time series $f(s_1, t)$ explains the most of the variance in the predictand data set $g(s,t)$ at lag τ, i.e. we find i for which $U(i)$ defined as

$$U(i) = \sum_j (\rho_{ij}^2 * \sum_t g^2(s_j, t+\tau)/n_t) \qquad (8.7)$$

is maximum. Having found s_1 we take $f(s_1, t)$ to be the first mode's time series of *both* f and g expansions, i.e. $f^{\text{explained}}(s, t) = a(s_1, s)f(s_1, t)$, and $g^{\text{explained}}(s, t+\tau) = b(s_1, s)f(s_1, t)$, where $a(s_1, s)$ is the regression coeffi-

---

[10] We use the name SVD, even though we agree with Zwiers and Von Storch (1999) that it is unfortunate that the name of the method is confused with a basic matrix operation; they suggest Maximum Covariance Analysis (MCA). MCA is not a great name either since a) we maximize SC and b) maximizing explained SC is not really the goal of forecasting g from f, see EOT2 discussion (p. 142–143)

cient to predict $f(s, t)$ from $f(s_1, t)$, and $b(s_1, s)$ is the regression coefficient to predict $g(s, t + \tau)$ from $f(s_1, t)$. The spatial patterns in (8.6) are thus: $e_1(s) = a(s_1, s)$ and $d_1(s) = b(s_1, s)$. Note that $b(s_1, s)$ is proportional to the correlation defined in Equation (8.5) and used in Equation (8.7). We then seek the second point in the once reduced data sets $f^{\text{reduced}}(s, t) = f(s, t) - a(s_1, s)f(s_1, t)$, and $g^{\text{reduced}}(s, t + \tau) = g(s, t + \tau) - b(s_1, s)f(s_1, t)$, to find s_2, etc. This procedure has many of the properties of CCA, specifically the identities (CCA-1), (CCA-2) and (CCA-3/3a), the latter with cor(m)=1 for all modes. (Oddly, EOT2 actually "beats" CCA on producing the highest correlation between the time series.) EOT2 has at least two notions of relevance: the EV in data set f, and the EV in data set g. The latter is what is maximized, albeit under the constraint that we use a single time series of f at one point in space (rather than linear combinations of f at various points). There does not appear to be a particular need for the explained SC, after all the target of the prediction is EV in g.

Making a forecast of g is easy. For the first mode we need the latest observation of f at s_1, then multiply by $b(s_1, s)$. Subsequent modes are similar, but f has to be $m-1$ times reduced for the mth mode.

The reader will not be surprised that there is an "alternative" lagged covariance matrix given by

$$c^a_{ij} = \Sigma f(s, t_i)\, g(s, t_j + \tau)/n_s \qquad (8.5b)$$

where summation is in space. Here we consider inner products of maps of fields f and g at times t_i and $t_j + \tau$. At first sight this definition is possible only if the domain and grid points for f and g are the same. However, this discrepancy is resolved by first executing EOFs on f and g individually and thinking of s in (8.5b) as the mode number. We now pick the one f map at time t_i which maximizes the variance explained in g, an expression similar to (8.7) but reversing the roles of time and space. This single map then acts as $e_1(s)$ for f and $d_1(s)$ for g. There are two time series, which are regression coefficients $a(t_1, t_i)$ to predict $f(s, t_i)$ from $f(s, t_1)$ and $b(t_1, t_i)$ to predict $g(s, t_i + \tau)$ from $f(s, t_1)$. This alternative EOT2 route leads to the expansion (8.6) and (8.6a) with the properties (SVD-1) and (SVD-2) but not (SVD-3). The alternative EOT2 has again two notions of relevance, the EV in data set f, and the EV in data set g. The latter is not only what is maximized,[11] but is the purpose of the regression.

The two EOT versions that closely bracket CCA (regular EOT2) and SVD (alternative EOT2) come with either two maps and one time series (nearest CCA) or one map and two time series (nearest SVD). From this it appears that SVD is subject to more orthogonality constraints than CCA; after all (CCA-3) follows trivially when there is only one time series to begin with, but (SVD-3) does not follow automatically from having a single map ($d=e$).

---

[11] Under the constraint that we use maps of f at one point in time (rather than linear combinations at various times).

Note that when admitting too many modes CCA/SVD goes in the direction of multiple linear regression. Obviously, truncation is necessary for reaping the benefits of regression at the pattern level.

Much information about SVD and CCA can be found in Bretherton et al. (1992), Newman and Sardeshmukh (1997) and Zwiers and von Storch (1999). Wilks (1995) provides a good discussion of CCA.

CCA was not used much in meteorology until Barnett and Preisendorfer (1987). The main methodological twist in their paper is a prefiltering step where both f and g are truncated to just a few EOFs before calculating C. Moreover, the EOF associated time series are standardized, as in a version of the Mahalanobis norm (Stephenson 1997) The prefiltering greatly reduces CCA's susceptibility to noise. The prefiltering also makes the practical difference between SVD and CCA in many instances very small. Additionally Barnett and Preisendorfer (1987) applied their adjusted CCA to the seasonal forecast and had the predictor data set cover four antecedent seasons. This method and this particular predictor lay-out has been popularized by Barnston (1994) and his work found short-term climate prediction application on nearly all continents: Johansson et al. (1998) for Europe; Thiaw et al. (1999) for Africa; Hwong et al. (2001) for Korea; Shabbar and Barnston (1996) for Canada; He and Barnston (1996) for tropical Pacific Islands and Barnston and Smith (1996) for the whole globe. While SVD is often mentioned in one breath with CCA, and widely used in research (Waliser et al. 1999; Wu and Dickinson 2005) there appear to be far fewer real-time forecast applications based on SVD. CCA is also applied as a method to correct errors in GCM predictions (Smith and Livezey 1999; Tippett et al. 2005)

As a diagnostic tool SVD or CCA may be as difficult to use as EOF, i.e. the patterns in the predictor and predictand data set may or may not be revealing the underlying physics. Plenty of examples of patterns are found in Barnston (1994). Newman and Sardeshmukh (1997) show the failure (to a certain extent) of SVD to discover that vorticity and streamfunction are linear transforms of each other. Zwiers and Von Storch (1999) also provide several examples.

We spent some paragraphs explaining SVD, CCA, etc. because so much of the modern empirical work is along these lines. Regression on the pattern level is thought to take away the arbitrariness of correlating everything with everything. Although methodological details are hotly debated sometimes, the other choices may be more important than the exact method. For instance, which predictors, how far back in time, how many time levels, the domain for predictors and predictands, pre-filtering, truncation, etc. may be more important than the exact CCA vs SVD method. The CCA at CPC and CDC, identically the same method, often give conflicting tropical Pacific SST forecasts. While we presented the above material as a strictly separated predictor f and predictand g, keep in mind that the data sets may be combined, i.e. fields of the predictand at an earlier time may be appended to f in order to forecast g. CCA has been used at both CPC and CDC for real-time seasonal prediction; skill levels are at best (short lead JFM seasonal T&P) 0.3–0.35 correlation nationwide with regional variations that reflect the large impact of ENSO (Barnston 1994; Quan et al. 2005). The CCA modes suggest lesser influences from other tropical areas and mid-latitude oceans as well.

8.7.3   LIM, POP and Markov

Somewhat similar to CCA and SVD are the linear inverse model (LIM) and principal oscillation patterns (POP). The similarity is in the central role of the lagged covariance matrix as in (8.5), evaluated from data. However, both POP and LIM try to generalize the results for lag τ to all other lags by assuming an underlying theory. Following the Winkler et al. (2001) notation one may assume a linear model given by

$$\mathrm{d}\mathbf{x}/\mathrm{d}t = \mathbf{L}\mathbf{x} + \mathbf{R} \tag{8.8}$$

where $\mathbf{x}$ is the retained scales state vector, $\mathbf{L}$ is a linear operator and $\mathbf{R}$ is random forcing due to unresolved scales (possibly with structure in space). Vector $\mathbf{x}$ would for instance be a combination of data sets f and g. The solution to (8.8) is

$$\mathbf{x}(t + \tau) = \exp\left(\mathbf{L}\tau\right)\mathbf{x}(t) + \mathbf{R}', \tag{8.9}$$

where $\mathbf{R}'$ depends on the history of $\mathbf{R}$. The operator $\mathbf{L}$ can be determined from data at a chosen lag τ_0, i.e. we evaluate $\mathbf{C}$ for lag τ_0. $\mathbf{L}$ is given by $\mathbf{C}(\tau_0)\,\mathbf{C}^{-1}(\tau = 0)$, see Winkler et al. (2001) for detail. The forecast for any lag τ is given by the first term in (8.9). The forecasts for τ_0 would, everything else being the same, be close to CCA's. But an analytical flavor is added because time evolution is implied. Moreover, it is possible to calculate the eigenvectors of the asymmetric $\mathbf{L}$ once and for all; they are structures evolving in time, and ultimately damped. By knowing the projection of the current initial state onto the known eigenvectors of $\mathbf{L}$, the forecast can be made analytically and can be interrogated for diagnostic purposes, such as in deriving the optimal structure to produce an El Nino pattern 10 months later (Penland and Sardeshmukh 1995). This is similar to what we presented for CA (section 7.6), although CA has additional growth due to unstable normal modes.

Several examples of POP, including for MJO forecasts, are given in Zwiers and von Storch (1999). Winkler et al.'s (2001) application is in the week-2 forecast, while Penland pioneered LIM for seasonal SST forecasts, both in the Pacific (Penland and Magorian 1993) and Atlantic (Penland and Matrosova 1998). In all cases $\mathbf{C}$ is calculated from EOF truncated data, but the degree of truncation varies wildly.

A straightforward method has been presented in Xue et al. (2000). In this paper the discretized version of (8.8) is used: $\mathbf{x}(t + \tau) = \mathbf{C}(\tau)\,\mathbf{C}^{-1}\,(\tau = 0)$ $\mathbf{x}(t)$, i.e. given an initial state $\mathbf{x}(t)$ and $\mathbf{C}(\tau)\,\mathbf{C}^{-1}(\tau = 0)$ as determined from data, the forecast for lead τ can be made. No linear model is assumed, so the calculation has to be done for each τ seperately, and nothing connects the forecasts at two different τ, except to the extent the data suggest. No modes are calculated, neither eigenmodes of L (as in LIM/POP, see Equation 8.9), or $\mathbf{M}$ (CCA) nor singular vectors of $\mathbf{C}$ (as in SVD). This cuts down on interpretation. The problem is handled as multiple linear regression, however after extremely heavy truncation using extended EOF in the input data. Xue et al. (2000) use sea-level height, wind stress and SST to forecast the same (sea-level height, wind stress and SST) in the tropical Pacific which appears to be a wise choice, since the method has worked well in real time. They call their method a "Markov" (MRK) method. CCA, SVD, and LIM, POP and MRK have options in truncation both in preparing the input data, and in truncating the modes calculated from C, L, or $\mathbf{M}$.

8.8   Numerical methods

This chapter would not be complete without discussing global atmosphere-ocean models. Such models are the extension of NWP technology down into the ocean and consequently into processes with much longer time-scales. It is generally assumed that most of the predictability of the seasonal climate resides in factors external to the atmosphere, such as the ocean, followed at some distance by land surface memory effects. In recent years, fully coupled atmosphere–ocean–land models have been developed to the point where they could start to make contributions to the real-time seasonal forecast (Demeter project, May 2005 Tellus 57A#3 issue; Saha et al. 2006 for the USA). This promise is realized only when plenty of hindcasts are available that allow both simple bias correction and more fancy calibration of pattern errors (Smith and Livezey 1999) and probability forecast adjustments. Raw model output can be very biased and needs post-processing. In practice the number of hindcasts that any institute can afford to run are limited by (a) computer power and (b) the period over which reliable initial states for land/ocean/atmosphere are available. Ocean analyses prior to 1981 may not be good enough as ICs or for verification. A 25-year period, multi-membered ensemble from each month is about the maximum that can be done (Saha et al. 2006) at the current time. The initial states for land are a concern as well, but here too there has been an enormous effort (JGR 2003 issue; Fan et al. 2006).

It is puzzling why comprehensive models, costing enormous resources, are only at par (roughly) with far simpler and empirical tools, and not much better. We come back to this conundrum in the final chapter. Saha et al. (2006) show the NCEP coupled model (named CFS) to be on par with CCA, CA, Markov, etc. in NINO34 forecasts, and close to being at par with simpler methods for US T&P seasonal prediction. Van Oldenborgh et al. (2003) have addressed the same question for the ECMWF model, and came to a similar conclusion.

8.9 Consolidation

Consolidation of multiple forecasts is necessary for a number of reasons. One can think of consolidation as the process of making the best possible official forecast out of many different forecast tools. This is a laudable goal for any weather service. While one may entertain the thought of consolidating subjectively; one very good reason for *objective consolidation* is that the supply of forecasts has become so large that no human (forecaster or user) is able to absorb the information in the available time, weigh their relative credibility, do justice to each component, and formulate the official forecast. And the problem is only getting worse. This is probably true for both short-and long-range forecasts, but here we address the latter. The idea of formal consolidation is at least as old as Thompson (1977), but development has been very slow.

We consider it self-evident that (a) one needs hindcasts in order to make an optimal consolidation and (b) the real-time forecasts need to be consistent in all respects with the hindcasts. But note that this practice has hardly been developed for dynamical models. CDC has undertaken a reforecast project (Hamill et al. 2004) for the week-2 forecast—this is a prototype of an atmosphere-only reforecast. The new coupled ocean–atmosphere model at NCEP (Saha et al. 2006), the CFS, has a hindcast amounting to 3500 years of integration. In the case of Hamill et al. (2004) the purpose was calibration of the model in its own right, not consolidation. The CFS also needs to be combined with other methods like CCA. Other literature or practices are often employing AMIP runs (Gates et al. 1999) for the hindcast of dynamical models (Peng et al. 2002), but this does not apply in the real-time setting. For statistical methods the idea of hindcasts comes more naturally, has been common since about 1990, and is fairly well developed

Consolidation is difficult for many small reasons, but for two large reasons in particular. The first is a stunning lack of data vis-à-vis the number of coefficients that need fitting. Think of consolidation as $CON = a \times A + b \times B + c \times C + \ldots$, where the capital letters refer to forecast tools, and the lower case coefficients need to be determined from hindcasts. Seasonal forecast tools like the CFS (Saha et al. 2006) have only a 25-year record of hindcasts. This may sound like a lot, but in terms of seasonal means there is very little independent material to tune coefficients. While some empirical methods like CCA (Barnston 1994) may have 50 (or potentially 100) years, the data set still falls short by orders of magnitude if there are many methods/models to be combined. This is especially true when A, B, C, ..., Z and all others are highly correlated. The second major difficulty is low-frequency climate change. The period 1981–2005 was much warmer over the USA than the 1960s and 1970s. This major source of forecast skill (when verified against an "old" normal) is accounted for in real-time operations at CPC by the Optimal Climate Normals (OCN) method (Huang et al. 1996). Dealing with this aspect retroactively on very limited data may be next to impossible.

By the nature of the consolidation methodology used one may see a stark distinction between two groups of consolidation activities, namely those working with point forecasts (it will be 66.0 °F tomorrow) and those working with probabilities (next winter a 50% chance of the upper tercile). The distinction is not as absolute as it may seem because in applying linear regression to N point forecasts, one automatically obtains the root-mean-square error (rmse), which, to a first approximation, serves as a standard deviation around the point forecast, i.e. linear regression results in a consolidated pdf as well. Whether the spread of forecasts, which varies on a case-by-case basis, can be used to improve upon the regression implied constant spread (based on rmse) remains an open question, but we note that Hamill et al. (2004) have concluded in the negative.

Consolidation is not new at NCEP. Leaving a long history of subjective consolidation aside, CPC has had a primitive consolidation of a few of its forecast tools. Van den Dool and Rukhovets (1994) designed a consolidation for the various members (at the time unequal members in terms of resolution, age) of the global model to support the 6–10 day forecast. Ever since Unger et al. (1996), CPC has presented the official NWS SST forecast (Nino3.4 only) as a consolidation of in-house tools (CCA, CA, MRK, coupled model); the exact method has evolved over time. In a sense "ensemble" CCA (Mo, 2003) is also a consolidation. At the IRI a Bayesian method and a canonical variate method (Barnston et al. 2003) are used to consolidate a large number of model forecasts from various centers around the world. The DEMETER project in Europe has been used to make a strong case in favor of multi-model ensembles (Hagedorn et al. 2005; Stephenson et al. 2005). Other research includes Kharin and Zwiers (2002) and Roulston and Smith (2003). CPC is working on consolidation as per ridge regression, see Appendix 2.

Consolidation should be better than the single best participating tool, but for all the work to be done to determine optimal weights, etc. the following should be kept in mind.

(1) The consolidation will not be much better than the best individual tool if there is little independent information provided by the other tools

(2) The consolidation might fail (in real time) if any of the hindcast data sets gives a flawed impression of skill or if we cannot execute real-time forecasts 100% consistent with the hindcasts. We do not have full control over these factors.

8.10   Other methods

In the above we have listed and described more or less all methods used operationally recently at institutions in the USA (principally CPC, IRI and CDC) concerned with short-term climate prediction. While it is totally impossible to be complete, we here mention briefly some other methods and applications. It is not our intention to make an inventory of a toolbox, just tools that have been developed into methods for short-term climate prediction.

Multiple linear regression (MLR) has been used for ages. In Germany, under the leadership of Franz Baur, a truly ambitious MLR program was pursued in the early part of the twentieth century. The effort was handicapped naturally by the shortness of records at that time. MLR is dangerous in that chance correlations can be believed too easily by the hopeful, and that appears to have been widely the case. The over-reaction to the early mistakes was one of doubting anything empirical. Even the correlations associated with the Southern Oscillation (Berlage 1957), now considered essentially correct and seminal to understand ENSO, were considered suspect and lost their appeal when attention and resources were absorbed into the emerging NWP effort. Barnett (1981) and others re-introduced MLR, and his work (and work by others, for instance Hastenrath and Greischar 1993) includes a serious attempt to test regression on independent data, under cross-validation, etc. so as to largely avoid the errors made in the Baur era. Variations on the EOF theme may lead to any number of MLR approaches (Vautard et al. 1999), and could include attempts to try nonlinear regression (Kravtsov et al. 2005).

There is a method called CLIPER, an acronym that suggests a combination of climatology and persistence, topics covered in sections 8.1 and 8.2. The history of CLIPER in hurricane prediction is to serve as a benchmark forecast to be improved upon by more complicated approaches. However, the CLIPER used in Landsea and Knaff (2000) and Knaff and Landsea (1997) for Nino34 prediction also has non-local predictors, and predictors at more than one previous time, see Sections 8.4 and 8.5. This is quite ambitious for a control method. CLIPER may well be among the better methods of forecasting Nino34.

Forecasts of PNA and NAO: the enormous attention for teleconnections and modes has given apparent reason to make PNA or NAO as such the predictand. A good example is Rodwell and Folland (2002) who concluded that a regression from May Atlantic SST to the NAO index in the next winter had more skill than SST predictors at other times. The quoted skill (in correlation) is about 0.4–0.5. A big problem with Atlantic predictors and the NAO are strong (inter)decadal variations. If relationships are strong at very low frequencies, there are very few degrees of freedom and claims of skill may have to be postponed until plentiful independent realizations are in. The UK Met Office has kept track of the performance of this scheme, see http://www.metoffice.com/ research/seasonal/regional/ nao/.

A comprehensive verification of NAO and PNA forecasts at time-scales ranging from daily to long lead seasonal and by many different methods is given by Johansson (2006).

There has recently been a shift in terms of predictands. Given the difficulty of forecasting precipitation, attention has shifted to space–time integrated variables such as runoff, river flow, snowpack and soil moisture. These quantities have longer time-scales, are easier to forecast (unless one invokes persistence as the control) than precipitation and are closer to application. Early examples include Cayan et al. (1995) who essentially made ENSO composites for snowpack in the western USA. Webster and Hoyos (2004) present a 10–30 day forecast scheme for major rivers in the South Asian subcontinent during the summer monsoon, based on a combination of wave-let analysis (Torrence and Compo 1998) and linear regression. Their predictors are chosen based on physical insights about the working of the Indian monsoon. Constructed analogue (Chapter 7) has been applied to soil moisture over the USA (Huang et al. 1996) as predictor, to predict temperature, precipitation, evaporation and soil moisture over the USA out to several months. While some skill was reported (Van den Dool et al. 2003) it should be noted that the high skill for soil moisture (0.6 correlation) does not beat persistence. Van den Dool et al. (2003) also contains a rock in the pond experiment where one places a round soil moisture anomaly somewhere in the USA, anomalies being zero everywhere else, then let CA determine the motion of the soil moisture anomaly in the next several months in response to rainfall and evaporation anomalies. Variations on spectral methods, particularly singular spectrum analysis and maximum entropy methods, can be traced via a review article by Ghil et al. (2002).

Checking the issues of the *Experimental Long Lead Forecast Bulletin* from its inception in 1992 to the present will provide access to a large array of methods experimented with by various authors. See http://www.iges.org/ ellfb/home.html.

There are good methods that have not been used in practice, although they could have been; this may be the accidental course of history up to this point. To delineate what is usable it may be best to list the requirement for any method to be usable. The main requirements are about the method itself, the reproducibility of the calculations, and the unambiguous format of forecasts that need to be verified. The methods should be documented for the purpose of peer review; it also allows a potential user to judge whether this method could be any good. It should be possible, at least in principle, to reproduce a calculation (objective tools) and also to interrogate a method: why does Method X forecast such and such anomaly? After the fact it should be possible to make an unambiguous verification. An additional verification requirement is nowadays that of certified skill over a long enough testing period, under carefully designed conditions (to avoid erroneous skill estimates). These requirements rule out subjective prediction, or mostly subjective prediction. There has to be a method in the madness. Keeping these requirements in mind we now discuss methods that we do not recommend.

8.11   Methods not used

Some 15 years ago the author, in his capacity of Chief of CPC's Prediction Branch, received a phone call from a *New York Times* reporter who first explained that a groundhog named Punxsutawney Phil had seen his shadow, so six more weeks of winter had to be expected. Then he asked: how does this compare to the official forecast. Suppressing a tendency to burst out in laughter, we politely answered the question. The reader of this book will probably agree immediately that sunshine in the early morning in one spot in Pennsylvania on February 2nd is an unlikely predictor of winter weather for the next six weeks. But how do we know so immediately? Based on the aforementioned requirements for usable methods the groundhog prediction fails on ambiguity (it is hard to verify the forecast as phrased), and also on method (this is unlikely to have skill). So we don't use Punxsutawney Phil in a professional environment. While it is impossible to prove categorically that any particular method is useless for any imaginable application we believe the same can be said about mapping the weather on the 12 days of Christmas onto the next 12 months, the hairiness of caterpillars in the fall, the size of beechnut crop, etc. This is diversion and entertainment. Somewhat harder to dismiss off hand are the (ensemble of) farmer's almanacs. Persistent questions forced university researchers Walsh and Allen (1984) to attempt verification. Occasionally, during a long warm spell, one can hear a weatherman on TV say: "things have to average out" so as to indicate that a compensating cold spell is unavoidable. This compensation idea is actually wrong. If it was true anti-persistence at a certain lag would prevail. A variant can be heard by economists commenting on the stock market: "what goes up must come down". These are truisms without any forecast skill (in meteorology) over a reasonable control. (Economic forecasts may come true when people believe the economist's truism and sell off their stock.)

A special topic is that of solar influences. While it makes perfect physical sense to assume that variations in solar output directly affect the atmosphere's temperature, we now know, from post-1979 satellite observations, that the solar constant varies only by about 0.1% over the course of the 11-year sunspot cycle. This may be detectable in global mean temperature, but is probably too small to be detected as a local effect. Possible amplification may exist via larger changes in ultraviolet radiation, and the stratospheric chemistry response. For more than a century before 1979 researchers have given the solar-weather connection a bad name by correlation everything with the 11-year cycle (known from sunspots). Even now anything solar as a predictor gets easily dismissed (over-reaction to past mistakes). There is a small probability this could change seriously (Lean and Rind 1999).

This section's title "Methods not used" does not quite cover the last example in this paragraph. In this case the method itself is fine, but its use and interpretation is debatable. In the distant past, the search of cycles (any cycles) was a prominent research activity. Even without identifying specific cycles as credible some methods involved Fourier analysis of very long time series, and the forecast for Δt ahead would be the extrapolation of all harmonics over Δt. This makes sense for periodic components (the atmospheric and oceanic tides), which we deleted from consideration from the outset as too easy to forecast (see start of this chapter). Fourier analysis followed by extrapolation to forecast anomalies in a chaotic atmosphere makes little sense (unless we have overlooked periodic components). Still this method is tried off and on, occasionally with modern twists such as EOFs, etc.

Appendix 1: Some practical space–time continuity requirements

Many researchers may feel that maximizing the skill of a method is what is needed most. This can be done by wise choices about the prediction method, the predictor/predictand data sets and a cautious approach about cross-validation. However, in practice there may be requirements that make the skill sub-optimal; these requirements are hard to deal with in research. We give a few examples.

1. Variation of K in OCN. In Section 8.3 we described how OCN is done. In principle this yields an optimal K value for each location, and every

rolling season. However, because the $U(K)$ may be a very flat function, and or have several minima the optimal K at nearby stations could be quite different. For instance when $K = 2$ in Washington DC and $K=18$ in Baltimore (60 km apart), very different OCN forecasts in real time could be the result. In order to avoid such irregularities on a national map, the function $U(K)$ is actually minimized by summing over all locations and all seasons (as well as years). $K=10$ for temperature was derived that way. For a party that has local interests only, and is not concerned about inconsistencies in space, the optimal K value (varying with season) may be better.

2. MLR inconsistencies. In a similar vain, multiple linear regression used at CPC (Unger, personal communication) leads to inconsistencies because the best prediction equations for two nearby predictands are derived without regard for each other, can be quite different in the choice of predictors and may lead to spatial variation in a real-time forecast that is impossible. It takes constraint on the part of the real-time forecaster to decide what to believe. Regression at the pattern level largely avoids these problems, and may be preferred for that reason alone.

3. Consolidation of methods. When the weights assigned to methods are irregular, it could happen that certain models are "coming in and going out" of the consolidation as a function of lead. This strikes the real time forecaster as an unlikely scenario. Measures should be taken to avoid that (by merging leads, points in space and neighboring initial conditions). Negative weights for forecasts also seem absurd. We assume that every reputable center does the best possible job and nobody has significant negative skill. (If they knew how to consistently give the anomaly opposite to observed they might as well give us the correct forecast.) Theoretically negative weights are possible but it is hard to sell a forecast for a positive anomaly in location X on the ground that world famous model A is forecasting a negative anomaly (and is always wrong; or gets a negative weight in a field of collinear forecasts).

Appendix 2: Consolidation by ridge regression

For US T&P we consider implementing a version of ridge regression as the consolidation method. It is assumed that this will function even with an 'overload' of participating methods and short data sets, and this is the key consideration. The solution will be kept sane by pulling it slightly in the direction of a simple-minded approach based on just the skill of each method. Since skill ought to be positive, negative weights should never be assigned to any method. Space dependence of weights may be possible to some degree.

Definitions

Let **A**, **B** and **C** be three[12] forecast methods with a hindcast history 1981–2003. **A** is shorthand for **A** (year, initial month, lead, space) or **A** (year, target month, lead, space). Stratification by month is customary, so **A** (y, l, s) suffices in the notation below, where y is 1981 to 2003, lead=1, 6(13), space (s) could be grid points NH (for example) or 102 super climate divisions in USA. The matching observations are **O** (y, l, s). The inner product is defined by:

$$\mathbf{AB} = \Sigma \mathbf{A}(y, l, s) \times \mathbf{B}(y, l, s) \qquad (8.A1)$$

where summation is over time y and (some or all of) space s. For simplicity we work with just three methods, but the derivation can be given for N methods.

In general we look for:

$$\text{CON(solidation)} = a^* \mathbf{A} + b^* \mathbf{B} + c^* \mathbf{C} \qquad (8.A2)$$

In a simple-minded solution a, b and c are proportional to the skill of methods **A**, **B** and **C**, i.e. proportional to **AO**, **BO** and **CO** (covariances), multiplied by 1/**AA**, etc. In that case $a + b + c$ probably needs an additional constraint like $\Sigma\ a+b+c=1$. a, b and c could be function of space, lead, initial (target) month. a, b and c should always be positive because we do not admit methods with negative skill (over the hindcast data set).

Full solution

While a simple-minded solution may be practical we actually seek the full optimal solution, taking into account both skill by methods and "collinearity" among methods:

$$\begin{bmatrix} \mathbf{AA} & \mathbf{AB} & \mathbf{AC} \\ \mathbf{BA} & \mathbf{BB} & \mathbf{BC} \\ \mathbf{CA} & \mathbf{CB} & \mathbf{CC} \end{bmatrix} \times \begin{bmatrix} a \\ b \\ c \end{bmatrix} = \begin{bmatrix} \mathbf{AO} \\ \mathbf{BO} \\ \mathbf{CO} \end{bmatrix}. \qquad (8.A3)$$

If collinearity were zero, note $a = $ **AO/AA**, etc, the simple-minded solution. Also note there is no constraint on $\Sigma a + b + c$. The full solution takes collinearity of methods into account, i.e. if **A** and **B** always give the same information they have to share the weight; they do not both get a high weight. The measure of collinearity is given by the strength of the off-diagonal elements, relative to the main diagonal (**AA**, **BB** and **CC**).

Including collinearity is essential for the full solution, but problems arise when the collinearity is too large, or when there is not enough data to estimate the collinearity accurately. In either case the solution (a, b, c) to Equation (8.A3) may be unstable. In consolidation of seasonal forecasts there is *Far* too little data to determine $a, b, c \ldots, z$, given the number of participating methods (quickly increasing).

---

[12] Three is just an example. Any number will do for explaining the process.

Ridge regression is an amendment to the full solution to address this problem. One can stabilize the solution by adding small positive constants to the main diagonal of the matrix. This changes (8.A3) to (8.A4).

$$\begin{array}{ccc} \text{Matrix} & \times \text{ vector} = & \text{vector} \end{array}$$

$$\begin{bmatrix} \mathbf{AA}+\epsilon^2 & \mathbf{AB} & \mathbf{AC} \\ \mathbf{BA} & \mathbf{BB}+\epsilon^2 & \mathbf{BC} \\ \mathbf{CA} & \mathbf{CB} & \mathbf{CC}+\epsilon^2 \end{bmatrix} \times \begin{bmatrix} a \\ b \\ c \end{bmatrix} = \begin{bmatrix} \mathbf{AO} \\ \mathbf{BO} \\ \mathbf{CO} \end{bmatrix}. \qquad (8.\text{A}4)$$

Even very small ϵ^2 can stabilize the worst possible matrices. Adding ϵ^2 to the main diagonal plays down the role of collinearity ever so slightly, and drives the solution very slightly in the direction (but not exactly so) of a simple-minded solution. A second layer of amplitude adjustment may be needed.

About ridging

The oldest reference on ridging in the English literature is Tikhonov (1977) in translation, but this method may have been known since 1950 in Russia. The basic idea is to find a reasonable solution where there are more unknowns than equations. Nominally we have three equations and three unknowns in (8.A3), but when collinearity is too large there may be effectively fewer than three equations. While (8.A3) minimizes the rms difference between O and CON on a given data set, the ridge regression minimizes simultaneously $\sum a \times a + b \times b + c \times c$. Ridging does this very effectively. The situation we encounter in consolidation is similar to data assimilation (Gandin 1965), where redundancy (collinearity) among observations to be assimilated is large. In the data assimilation context ϵ^2 relates to the (assumed) error in the observations. One could even embrace the situation as follows.

- Truncate forecast(obs) in EOF space.
- Now determine AA from filtered data
- Add ϵ^2 which is related to variance of unresolved EOFs.
- Controlled use of noise: off-diagonal elements unchanged.
- Solve the system.

In this context ϵ^2 has a real meaning, namely the variance of the unresolved EOFs.

9 The Practice of Short-Term Climate Prediction

While previous chapters were about methods and their formal backgrounds, we here present a description of the process of making a forecast and the protocol surrounding it. A look in the kitchen. It is difficult to find literature on the subject, presumably because a real-time forecast is not a research project and potential authors (the forecasters) work in an ever-changing environment and may never feel the time is right to write an overview of what they are doing. Moreover, it may be very difficult to describe real-time forecasts and present a complete picture. Nearly all of the material presented here specifically applies to the seasonal prediction made at the NWS in the USA, but should be relevant elsewhere.

A real-time operational forecast setting lacks the logic and methodical approach one should strive for in science. This is for many reasons. There is pressure, time schedules are to be met, input data sets could be missing or incorrect, and one can feel the suspense, excitement and disappointment associated with a forecast in real time. There are habits that are carried over from years past—forecasters are partly set in their ways or find it difficult to make major changes in mid-stream. The interaction with the user influences the forecast, and/or the way the information is conveyed. Psychology enters the forecast. Assumptions about what users want or understand do play a role. Generally speaking a forecast is thus a mix of what is scientifically possible on the one hand and what is presumably useful to the customer on the other. The CPC/NWS forecasts are moreover for the general user, not one user specifically. Users for short-term climate forecasts range from the highly sophisticated (energy traders, selling of weather derivatives, hydrologists) via the (wo)man in the street to entertainment.

The seasonal forecast has been around a long time in the USA. Jerome Namias started in-house seasonal forecasts at the NWS in 1958. After 15 years of testing, his successor Donald Gilman made the step to public release in 1973. The seasonal forecast had been preceded by a monthly forecast (starting in the early 1950s) which in turn was preceded by a 5 day mean forecast for days 2 through 6 (well before NWP played a role), a project that started around 1940 with several collaborators at MIT (Rossby, H. Willett, J. Namias). Some of the attributes of today's seasonal forecast, for instance the use of a three-class system, date back to these early efforts around 1940 (Rossby 1941). For a few more historical notes, see the appendix.

In this chapter we discuss a number of issues, specifically the rationale for time averaging, the lay-out and format of the forecast, the (in)famous three-class system, what is forecast and by which methods, a priori and a posteriori skill, hindcasts, the role of trends, etc.

9.1   On the seasonal mean

Why are we forecasting the seasonal mean, an average over about 90 days? Users, when asked, may express a desire for daily forecasts out to infinity, but here the limits set by state of the art science prevail over user desires. It is impossible as of now to forecast, with skill, day-by-day weather beyond one or two weeks. Figure 9.1 is an example of 500 mb height verification out to 30 days in a 5-year forecast data set produced retroactively by a 2002 NCEP global NWP model (Jae Schemm, private communication). Beyond two weeks the correlation of daily forecasts with verifying analyses is small (<0.20), even in the leading modes NAO and PNA, but not completely zero either. Assuming the remaining correlation is worthwhile to some users a time mean is taken as a filter to amplify the signal to noise (SN) ratio; note that many verification measures relate to the SN ratio (Compo and Sardeshmukh 2004). The signal is defined here as the predictable part of the

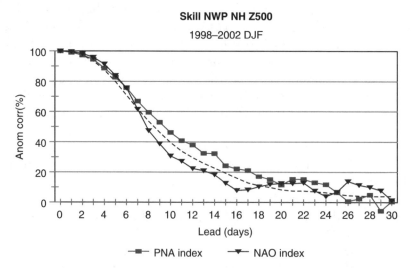

Figure 9.1   Typical decrease of anomaly correlation of daily Z500 forecasts as a function of lead time (black dashes). From a practical point of view skill is too low for day-by-day forecasting after 5, 10 or 15 days (depending on criterion). However, the correlation remains slightly positive at longer leads (even after 30 days). From Jae Schemm's 5-year calibration data set (NCEP-GFS model)

weather, while the rest is noise. Forecasting the seasonal mean over day 15–105 ahead of time is thus somewhat of an admission up front that skill is inherently limited in that range. For similar reasons NWS's CPC has had a 6–10 day averaged forecast (since 1978) and a week-2 forecast (since 1997) where low skill is addressed by taking a 5- and 7-day mean, respectively. The lower the correlation (but still positive), the stronger the averaging one would need to reduce the noise sufficiently. But one also needs to be sure the signal does not get harmed by the averaging. (A time mean over the first week of NWP forecasts would be unwise because the signal, well forecast early on and time varying, is harmed by taking a time mean.) A time mean in a situation of nearly constant signal (i.e. constant with lead time) is, in purpose, comparable to taking the mean of a modern ensemble (Tracton and Kalnay 1993); the purpose is to improve skill by some measure.

The transition from week-2 to a season in terms of averaging length is rather abrupt. Indeed a monthly (mean) forecast in between week-2 and seasonal may seem advisable. Currently, the intraseasonal forecast (say a monthly mean from day 15 to day 45) is still very difficult and has low skill, lower than the seasonal mean at longer lead. At ultra-long leads one could consider time averages longer than a seasonal mean, but here the user's needs prevail. Few users would be served by an annual mean forecast, even if it had some skill. For similar reasons prediction of spatially averaged quantities are rarely considered practical (all weather is "local"), even though in research "all-India" rainfall has been the target of prediction (Mooley et al. 1986). The seasonal mean is probably the longest time average one can afford without mixing wildly different winter and summer climates.[1]

9.2 Lay-out of the forecasts

Figure 9.2 shows a lay-out of the forecasts at NWS; time progresses towards the right. The day-by-day short-range forecasts (day 1–7, not discussed here) are followed by the 6–10day/week-2 forecasts which are already statistical in nature in that the target is a time mean, and the format used is probabilistic (O'Lenic and Handell 2004). Through week 2, i.e. through day 14, the basis of the forecast is almost entirely in NWP, with its suite of ensembles (Tracton and Kalnay 1993). The shortest seasonal forecast is applied to day 15–105 averaged, its "lead time" is two weeks. (The lead time is defined as the amount of time between the moment a forecast is issued and the first moment of validity.) Then 12 more rolling seasons

---

[1] OCN is the exception. While OCN is a 10 or 15 year average, it is applied only to a certain target season.

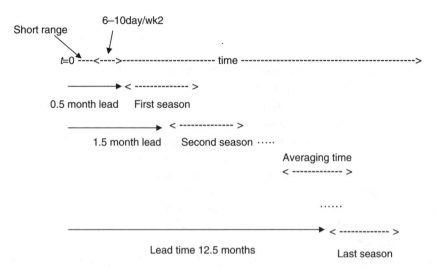

Figure 9.2 A lay-out of the seasonal forecast, showing the averaging time, and the lead time (in red). Rolling seasonal means at leads of 2 weeks to 12.5 months leads are being forecast. The forecast for the first season starts where the short-range and week 2 forecasts leave off.

follow, each increasing the lead time by one month, out to 12.5 months, see Figure 9.2 for a schematic. For example in mid-November 2005 seasonal forecasts were prepared covering DJF2005/06 (lead 0.5 month), JFM2006 (lead 1.5 month), ..., through DJF2006/07 (lead 12.5 month). This suite of forecasts is released every third Thursday of the month. The DJF2005/06 forecast was first issued as a 12.5 month lead forecast in November 2004.

At CPC the expression "long-lead" is used. This refers to the fact that even the first seasonal forecast starts beyond week 2. In the past, prior to 1995, the seasonal forecast started at lead zero, i.e. started its validity immediately after release. The wisdom of having (or not) a zero lead seasonal and monthly forecast continues to be debated among users, forecasters and researchers.

9.3   Time-scales in the seasonal forecast

While seasonal sounds like 90 days, the time-scales that need to be kept in mind are several:

(a) averaging time;
(b) lead time; and
(c) time-scale of physical processes that contribute to skill.

We again refer to the lay-out of the seasonal forecast in Figure 9.2 for explaining (a) and (b). The averaging time (our choice) is the easiest: 3

months. The *lead time* is the amount of time between the moment a forecast is issued and the first moment of validity. For instance a DJF forecast issued on November 15 has a 0.5 month lead. Lead time and averaging times are choices and could be altered if necessary. The time-scale (c), beyond anybody's control, is related to physical processes in the geophysical system that can be tapped to make a forecast with some skill. Among the most important processes we have mentioned in previous chapters are ENSO (time-scale many months to a few years), low-frequency variations and trends (multi-year, decades or more time-scale) and soil moisture effects (a few months). In principle a full spectrum of time-scales is or could be relevant to the seasonal mean climate.[2] It may come as a surprise to some that multi-decadal low-frequency variations (e.g. OCN, see Section 8.3) should play a big role in the seasonal forecast. This is a reflection of not only the strength of such trends, see sections below, but also the lack of skill at the shorter time-scales of the spectrum.

9.4   Which elements are forecast, and by which methods?

Table 9.1 shows the elements being forecast officially at CPC and the methods used to accomplish this. The *elements* (left to right) consist of seasonal mean T&P, the sea-surface temperature and the continental soil moisture (w). For T&P the official forecasts are restricted to the USA (including Alaska and Hawaii), even if several tools are for the whole world. The SST is forecast for the whole world ocean, but the only official NWS SST forecast refers to Nino34. SST plays a role comparable to Z500 in short-range weather prediction, i.e. its skillful forecast is important because simultaneously occurring surface weather can be derived from it for the locale of interest. w is part of a pseudo-official forecast only in that the Drought Outlook is largely based on and verified against w. Many other elements are forecast by tools, but have no official status. Of the *methods* listed in Table 9.1, CCA, OCN, CA, ENSO composites, CFS, MRK, MLR and consolidation have all been discussed in Chapter 8. The CCA, OCN and CFS are the standard tools for US T&P used every month, while the CA for soil moisture and ENSO composites are examples of tools of opportunity, used only during the warm half of the year (soil moisture) and during ENSO winters (composites are invoked when forecasts for Nino34 indicate a warm or cold event). Any of the tools can be accessed in real time via http://www.cpc.ncep.noaa.gov/products/ predictions/90day/tools/briefing/. The "other models" mentioned in Table 9.1 are imported from institutions

---

[2] Time-scales less than 90 days are also present in a 90-day mean (Madden 1976), and mainly to the detriment of skill.

Table 9.1. Overview of elements (left to right) and methods (top to bottom) that play a role in CPC's seasonal forecast.

Method\Element	US-T	US-P	SST	Soil Moisture (w)	References
CCA	×	×	×		Barnston (1994)
OCN	×	×			Huang et al. (1996)
CFS	×	×	×		Saha et al. (2006)
CA-SST			×		Van den Dool and Barnston (1994)
CA-w	×	×		×	Van den Dool et al. (2003)
ENSO Composites	×	×			
Other Models	×	×	×		
Markov (MRK)			×		Xue et al. (2000)
Consolidation	×	×	×	×	
MLR	×	×			Unger (1996)

outside NCEP, subject to timeliness, a priori verification and other operational protocols, etc. Including all members in each institution's ensemble, the forecaster has access to order 100 different forecasts, a formidable task for any human being. The "consolidation" method was either (certainly in the past) a subjective process of combining all information, or (in the future) a largely objective combination of tools as described in Section 8.9.

Table 9.1 also shows some historical carry over and "accidental" aspects. Instead of CCA for instance, a number of similar methods could have been developed, see Section 8.7. This circumstance often depends on the preference and interest of personnel at a particular time. If a method listed in Table 9.1 is used to forecast only one or two elements, that does not imply it could not be used also to forecast the other elements, just that the research and development was not done.

CCA as a method was discussed in Chapter 8. The lay-out of the predictors of the original CCA at CPC is such that global SST and 700 mb height during the last four non-overlapping seasons are truncated by EOFs and compressed into a low-dimensional predictor vector. The predictand is also heavily truncated before the CCA is done. The truncated predictand at an earlier time is part of the predictor. More recently, a variant called ensemble CCA (Mo 2003), ECCA, has been added, but for the first lead only. This ECCA is based on upper level velocity potential, soil moisture, etc. but at one antecedent time level only. The MLR at CPC (Unger 1996) was developed to be a methodological alternative to CCA but with identically the same predictor–predictand lay-out. Soil moisture was added as an additional predictor.

In addition to the tools mentioned, there may be more informal aids. In fact the forecaster has a mental checklist that includes "local" effects (especially SST anomalies along the south California coast), short-term persistence, the very latest on Nino34 (and an adjusted larger or smaller

role for ENSO composites), opinions expressed in a monthly phone conference (often backed up by a researcher on the outside running a variation
of an accepted tool), etc. In addition to individual tools, consolidated
renditions of any two or three tools are available, for example ENSO
composites combined with trend (Higgins et al. 2004), CCA and OCN
combined, etc.

9.5 Expressing uncertainty

Because of its limited skill, it is important to express uncertainty for the
seasonal forecast. Whatever little skill is available should not get lost in
translation. This was recognized early (Gilman 1985), well before NWP
had ensembles, and long before probability forecasts were an acceptable
wide-spread practice in NWS (it still isn't!). It is apparently an article of
faith that uncertainty shall be expressed through a probability forecast.
One may think here of error bars (the standard deviation of a Gaussian
distribution around the point value), or a complete probability density
function (pdf). As shown in the example in Figure 9.3, the forecast is
basically thought of as a statement that Nature will draw a realization
from the conditional pdf (cpdf; dashed). (Note that Figure 9.3 features
pdfs of seasonal mean values, not pdfs of daily values during the season.)
The word conditional refers to a pdf subject to the initial condition and all
that is knowable about the future at that time in that circumstance. If there is

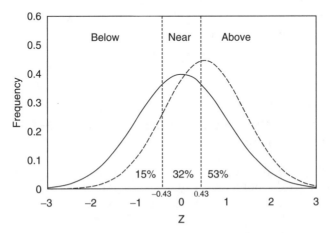

Figure 9.3 The climatological pdf (full) and a conditional pdf (dashed). The integral under
both curves is the same, but due to a predictable signal the dashed curve is both shifted and
narrowed. In the example the predictor–predictand correlation is 0.5 and the predictor value
is +1. This gives a shift in the mean of +0.5, and the standard deviation of the conditional
distribution is reduced to 0.866. Units are in standard deviations (x-axis). The dashed vertical
lines at ± 0.4308 separate a Gaussian distribution into three equal parts.

no skill, and the forecaster understands this correctly, the forecast should match the climatological pdf (full black line; as determined for instance from data over a standard climatology era, such as 1971–2000). In the example in Figure 9.3 the forecast has a warmer pdf (if this is temperature) by a noticeable shift to the right and suggestive of skill both by the shift in the mean and a narrower and higher distribution. As is also conveyed, a positive point forecast (the median of the distribution) in a situation with skill (a +0.5 shift and narrowed pdf) does not rule out a (high) negative value, just that the probability has been reduced greatly. More on Figure 9.3 later.

We here present pdfs and probabilities as a means to express forecast uncertainty, but these concepts have a less than obvious intersection with the concepts weather and climate (prediction), see also the first footnote in Chapter 1. Diagnostically it makes perfect sense to define climate as the pdf, and weather as a single realization drawn from that pdf. When using cpdfs as a means to express forecast uncertainty it thus makes some (not perfect) sense to look upon a cpdf as a climate prediction, while deterministic single forecasts or any point forecast would be weather prediction. This nomenclature appears to be followed more or less these days. The logic is imperfect because the needs for pdfs occurs at much shorter leads for P than for T. Nobody has declared the probability for P for the next 12 hours a climate prediction.

While the shorter range forecasts may have largely escaped a formal probabilistic approach (temperatures in the low 50s), the longer range forecasts have excelled in following a precise probabilistic protocol. Main problem: will the public understand this? How to convey probability information (such as in Figure 9.3) understandably and correctly to a large audience is a subject of continuing discussion.[3] It seems obvious that many users would not appreciate a "complete" pdf, whether it is provided as a graph, analytical function or a detailed tabulation for each locale. A simplification is needed, as described directly below.

9.6 Simplifications of the probability forecast (the three classes)

Instead of a complete pdf, which would in principle be an analytical expression plus the numerical values of the parameters describing it, the following simplifications have been used for over 20 years.

---

[3] Short of providing a complete pdf, uncertainty has been conveyed in several other ways. Maximum temperatures in the lower 50s has a purposeful uncertainty, i.e. the forecaster would not dare to say 52.4°F. The use of explicit error bars is uncommon in meteorology, but the use of categories, and making categorical statements (temperature will be in the below normal class) is similar to 100% confidence bars (open ended on one side for the two outer classes).

(1) Three classes, or terciles, are used.[4] Based on the climatological pdf three classes can be defined, named Below Normal, Near Normal and Above Normal (B, N, and A). In Figure 9.3 the vertical dashed lines at $\pm$ 0.4308 (in standardized units) signify the tercile boundaries for a Gaussian distribution.[5]

(2) By integration the conditional probabilities for B, N and A can be determined. In the example in Figure 9.3, one finds 15%, 32% and 53%, respectively. Indeed the probability for an above normal outcome has increased noticeably, mainly at the expense of the other extreme. The probability for the N class changes surprisingly little, unless the cpdf shifts are considerably larger than the half standard deviation used in the example.

At this point the simplified probability forecast, at each locale, consists of three numbers (p_B p_N and p_A) which, since they add up to 100, could be given by two numbers per location, still a complex map, i.e. two maps collapsed into one. The desire to present information as a simple under-standable national map for public consumption (like any other weather map) forces another simplification:

(3) $p_N = E$ and $(p_B - E) = -(p_A - E)$, where E is the climatological prob-ability (1/3). Equivalently $p'_B = -p'_A$ and $p'_N = 0$, where p' is the probability anomaly.

The seasonal forecast maps issued by CPC show contours of p'_B or p'_A, whichever is positive, with E added back in.[6] In the absence of any skillful information about the future the forecast, labeled nowadays EC (equal chances), would be 1/3, 1/3, 1/3.

For the advanced user the (more) complete pdf can be accessed in tabular or graphical form for many locations in the US, and none of the above simplification is used. See http://www.cpc.ncep.noaa. gov/pacdir/NFORdir/HOME3.shtml for real-time examples.

The use of three classes, while meant as a simplification, also creates an array of problems and questions.

(1) Why three classes, and why three *equal* classes? Clearly, a large number of classes is in the limit the same as a complete pdf, so, in order to simplify we need to reduce to just a few classes. Low skill also argues in favor of only very few classes. An odd number of classes would seem preferable as it

---

[4] The three-class system for categorical forecasts is at least 65 years old (Rossby 1941). Three-class probabilities were introduced in 1982, see Gilman (1985).

[5] The discussion is easiest for a Gaussian distribution, but three classes can be defined for nearly any distribution. CPC uses a two-parameter gamma distribution for P.

[6] For nearly 10 years we made maps of probability anomalies, but went back to full probabilities recently at user request.

leaves the neutral middle, the maximum of the pdf, as one entity. Some organizations have, however, used two classes, cutting the pdf in two parts right at the median, thus forecasting only the sign of the anomaly.[7] Three classes is thus the lowest number of classes that, in our opinon, makes sense as a simplification of the full cpdf. Until 1995 CPC used to have three classes based on a 30/40/30 climatological distribution. The wider N class was implemented to combat the lack of skill in the N class, a, by now, well understood problem (Van den Dool and Toth 1991), but the unequal classes always raised questions with users. In 1995 we went to three "equal" classes. Here is another reason why equal classes simplify: the notion "equal chances" would make no sense if the (30,40,30) classes are used. Indeed, in the past we have used various other symbols for EC: I (indeterminate), and CP and CL (both meaning climatological probabilities).

(2) The three classes have become so much the public face of the forecast that many people, even insiders, appear to have forgotten that it is meant as a simplification.

(3) Another mystery to many is "the event". Probability forecasts tend to be, in statistical parlance, for "an event". When the event is rainfall (or being hit by lightning) most people understand the concept, because it rains or it does not rain. The 50–50 concept as it relates to the flip of a coin is also widely understood. However, when the event is temperature falling in one of three terciles, the abstraction level is suddenly a challenge and "the event" somewhat mysterious. Explaining the situation with dice, or, by abstraction, a three-sided die might help. One could say that if Nature throws the loaded three-sided die in the example in Figure 9.3 an infinity of times, the B, N and A sides would appear 15, 32 and 58% of the time.[8] Clarification by invoking concepts in gambling (the odds[9]), flip of a coin, while highly applicable, is not uniformly appreciated by management at all times, because it suggests a non-serious activity.

(4) With modest probability shifts it frequently happens that the most favored class has less than 50% chance being categorically correct. By implication the favored class is more likely wrong than right. This causes bewilderment. (For these users the two-class system may be better.)

(5) There are in general negative connotations associated with any probability forecast. To many it seems as though we are seeking a formulation to never be completely wrong. By the same token a probability forecast is never a complete hit, unless one places 100% probability in the correct bin.

---

[7] At times CPC has had a two-class version of the three-class forecast.

[8] Never mind that Nature does it only once.

[9] For many years we were not allowed to use the word "odds".

9.7 Format of the forecast

Figure 9.4 shows an example of a set of forecasts released to the public in the middle of March 2006. These are the 0.5 month lead seasonal forecasts for AMJ 2006. Temperature is on the left, precipitation on the right. There are basically four options the forecasters have at their disposal to fill in these maps:

(a) A shift of probabilities towards above median, as in Figure 9.3, and as shown in Figure 9.4 in much of the south-western USA for AMJ 2006 for T. The contours, 33, 40, 50, etc. (with 1/3 subtracted) indicate how much the probability shift to the above median tercile amounted to. A suggestive color is used: orange-red (green) for above median[10] T (P).

(b) The same as (a) but now a shift of probabilities towards below median. Here the colors are blue for T (as in north-western USA) and brown for P (southern states).

(c) Equal chances (EC); $p_B = p_N = p_A = 1/3$. This would be areas left blank where no single tool has non-zero a priori skill, or signals by various tools with alleged skill are in conflict. EC is an informed "we don't know".

(d) An option (not used in Figure 9.4) for enhanced probabilities of the Near Normal class. Occasionally we give an N5, meaning that we borrow 2.5% from both extremes to make the distribution higher and narrower, but no shift. This would happen in an area with very high skill (in general) but a low signal on a particular occasion,[11] and also if two high-skill tools give opposite forecasts. The N option is rarely used because skill is so low for Near Normal (Van den Dool and Toth 1991; Kharin and Zwiers 2002). This is caused mainly by the unfavorable ratio of the width of the class to the rmse (the error bars); in a dry climate one has the same problem with the B class on P being very narrow (and one shower kills a forecast for B).

Under options (a), (b) and (d) only positive contours are shown. The reader is supposed to know the implied negative probability anomalies for the other classes, as described in here.

We have to accept this reality: Many users, and even some insiders, will simply look at the color, forget the contours, the pdf and the assumptions, and convert the map into categorical forecasts. Orange is thus above normal, green is above median, etc. and forecasts will be judged categorically.

The colored areas on the maps are sometimes referred to as non-EC.

---

[10] We use the notion median instead of mean because the precipitation is skewed. Median and mean is the same, or very nearly so, for seasonal T.

[11] For instance, if Nino34 correlates very highly with seasonal temperature at a locale of interest, then the chances of either extreme class to occur are reduced in a neutral year.

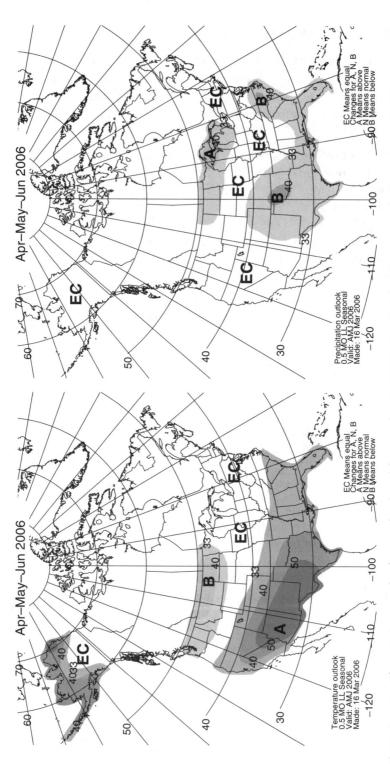

Figure 9.4 (see Plate 11) An example of a recently released forecasts for AMJ 2006, *T* on the left and *P* on the right. Contours are absolute probabilities at 33%, 40, 50 and 60%. The color and the letters A or B indicate the shift in probability towards Above (A, T red, P green) or Below (B, T blue, P brown). White areas labeled EC have climatological probabilities for all three classes.

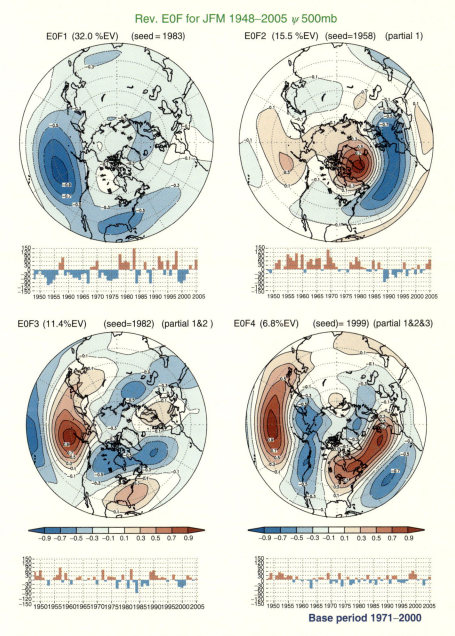

Rev. EOF for JFM 1948–2005 ψ 500mb

EOF1 (32.0 %EV)   (seed = 1983)

EOF2 (15.5 %EV) (seed=1958)   (partial 1)

EOF3 (11.4%EV)   (seed=1982)   (partial 1&2)

EOF4 (6.8%EV)   (seed)= 1999) (partial 1&2&3)

Base period 1971–2000

Plate 9 (corresponds to Fig 5.9): Display of four leading EOFs for seasonal (JFM) mean 500 mb streamfunction. Shown are the maps and the time series. A post-processing is applied, see Appendix 1, such that the physical units ($10^5 m^2/s$) are in the time series, and the maps have norm=1. Contours every 0.2, starting contours ± 0.1. Data source: NCEP Global Reanalysis. Period 1948–2005. Domain 20°N–90°N

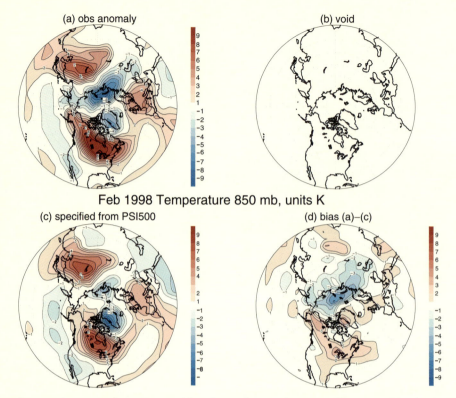

(a) obs anomaly

(b) void

Feb 1998 Temperature 850 mb, units K

(c) specified from PSI500

(d) bias (a)–(c)

Plate 10 (corresponds to Fig 7.5): The observed anomaly in monthly mean 850 mb temperature (upper left), the specified 850 mb temperature by the constructed analogue in (c) and the difference of (c) and (a) in (d). Contour interval 1 K. Results for February 1998. Map (b) is intentionally left void.

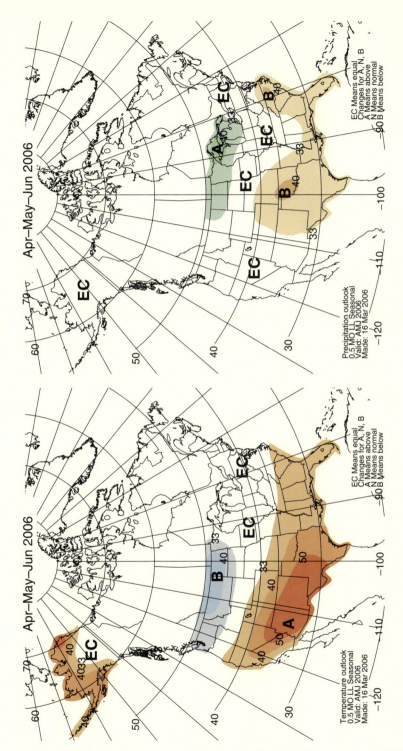

Plate 11 (corresponds to Fig 9.4): An example of a recently released forecasts for AMJ 2006, *T* on the left and *P* on the right. Contours are absolute probabilities at 33%, 40, 50 and 60%. The color and the letters A or B indicate the shift in probability towards Above (A, T red, P green) or Below (B, T blue, P brown). White areas labeled EC have climatological probabilities for all three classes.

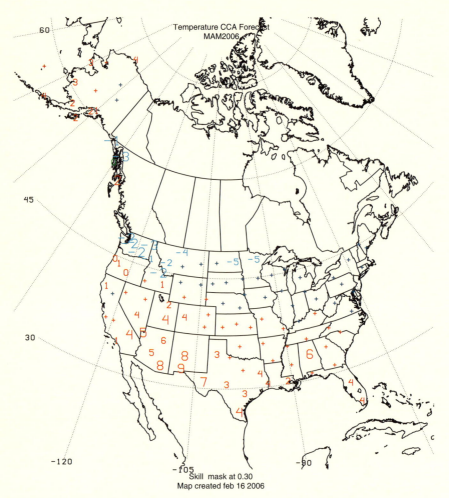

Plate 12 (corresponds to Fig 9.6): The looks of a tool used to make the seasonal prediction. The tool is the CCA. The units are standard deviations multiplied by 10. Red (blue) values are positive (negative) anomalies. The size of the numeral indicates the level of a priori skill. At the locations, indicated by only the red or blue plus sign, the a priori skill is below the 0.3 correlation, and no forecast value is shown.

The opposite of option (d), a wider pdf with reduced (enhanced) probability for N (outer classes) is technically possible but not practiced in the official forecast.

9.8   The official forecast

Table 9.1 is just a listing of 'tools' to make the official forecast. But how is the official made? It is convenient to think of the official forecast as a (linear) combination of the tools, e.g.

$$\text{OFF} = aA + bB + ,\ldots, + zZ, \tag{9.1}$$

where the capitals refer to methods (CCA, CFS, etc.), and the lower case coefficients depend on the skill of each method and the collinearity among them, see Section 8.9. Assuming we know the skill of the forecast (from many hindcasts, see next section) we need to convert Equation (9.1), a point forecast, to a probability forecast. For the three-class system this can be done directly, in simplest form, as per figures like Figure 9.5, which show the probability anomalies (p') for the extreme class (say the A class) as a function of (a) a priori skill (expressed as a correlation labeled R), and (b) the departure of the point forecast from climatology (=shift of the cpdf, labelled F). Figure 9.5 was prepared by David Unger. As expected, probability anomalies increase with both the correlation (R, in the vertical) and the strength of the point forecast (F, in the horizontal). While this is qualitatively quite obvious, Figure 9.5 provides a quantitative conversion depending on two knowable factors. These two factors, R and F, are not totally independent of course. In a situation of zero R, the anomaly point forecast should have been damped to zero. But for a modest non-zero R of say 0.5, the value of F, when extreme, can make a large difference in the probability. The same graph can be used for both A and B (but note the asymmetry relative to $F=0$), and the remainder for N then follows. For large R and F the p' for the extreme class is more than E (1/3); at this point one of the simplifications ($p'_A = -p'_B$) we described in Section 9.6 can no longer be applied and one would need to rob points from the N class as well as from the opposite extreme.

Because the suite of 13 seasonal forecasts is made each month, a certain target season at lead τ (except the very last $\tau = 12.5$ months) already has last month's official forecast at lead 1.5 as first guess. This way corporate expertise is handed down for 12 months in a row until the final opinion at the shortest lead = 0.5 is issued. So (9.1) could be written:

$$\text{OFF} = \text{first guess} + aA + bB + ,\ldots, + zZ + \text{subjectivity}. \tag{9.1a}$$

The subjectivity should be kept to a minimum.

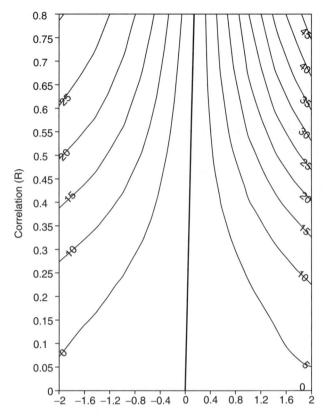

Figure 9.5   The probability shift (contours), relative to $100 \times 1/3$rd, in the above normal class as a function of a priori correlation (R, y-axis) and the standardized forecast of the predictand (F, x-axis). The result is based on shifting and or shrinking (the standard deviation of) a Gaussian distribution. The probability anomaly increases with both F and R. (Source: Dave Unger.)

9.9   Verification 1: a priori skill and hindcasts

Verification has already been mentioned several times, and indeed, verification is part and parcel of any credible forecast operation. In short-term climate prediction two kinds of verification exist. The obvious is the a-posteriori verification; after the fact one wants to determine the skill of (a particular set of) forecasts (see Section 9.10). Less obvious is the so-called a priori verification. The latter was developed to address situations with modest skill and/or situations where forecasts are issued infrequently. In these cases it is of paramount importance to give the user a sense of how much faith to put in the forecast. A forecast without some sense of a priori skill can do more harm than good. In the short-range weather forecast there are very many forecasts in quick succession in real time that may give the user an impression of the skill level, which can then be mentally applied to the next forecast by the user, but in short-term climate prediction there are only a few independent forecasts per year. No-one remembers far enough back to accumulate sufficient statistics of real-time forecasts for, say, DJF. One alternative is to evaluate a (large enough) set of retroactive forecasts (also called hindcasts) which mimic the real-time situation as faithfully as possible. Around 1990 this approach was first in place for several individual statistical objective tools that aided in the seasonal forecast. The need for hindcasts has increased exponentially since then because the number of tools is increasing very quickly. The relative weights of forecast tools in Equation (9.1) cannot be determined unless there are (enough) hindcasts to base them on. This is especially so when collinearity among tools is large, see the discussion on consolidation in Section 8.9. Hindcasts are also needed to bias correct, calibrate (probabilities) and verify each tool in its own right. Hindcasts can only be made for objective tools. Subjective forecasts and even the official forecasts cannot be credibly rerun over the last 25 or 50 years.

For statistical-empirical tools developing a set of hindcasts is easy, in principle, and can be done for a period covering the length of the data sets involved (~ 50 years). To make multi-membered hindcasts for a dynamical coupled ocean–atmosphere model the investment is much larger, and demands on CPU very high. Moreover, reliable initial conditions for the global land and ocean as required by dynamical models may not (yet) exist or may always be impossible given data scarcity, especially before 1980 (ocean). Because a consolidation is most easily based on the common period of the hindcasts, the operations at CPC and NCEP uses the period 1981–present for the hindcasts (even if normals are 1971–2000). The need for hindcasts has increased the need for observations, the need for recovery of nearly forgotten observations and the need for state of the art global reanalyses of which Kalnay et al. (1996) and Kistler et al. (2001) was only the beginning.

Hindcasts, if affordable, have this major advantage. Every time a tool is changed a new set of hindcasts would be available and there is no need to wait months or years before making an assessment of the skill of the new tools in real time operations. This assumes a set of diagnostics and verification can be run instantly.

Even for statistical tools a hindcast data set cannot be obtained without some investment. Statistical tools may suffer from overfitting and give a too optimistic view of skill to be expected in subsequent real-time forecasts. In view of a general impatience among clientele and funding agents, the old way of making forecast in real time and waiting until one has a sufficiently sized data set for evaluation will not do any more. Hindcasts are thus the approach of choice. To combat overfitting in hindcasts, "cross-validation" has been invented. In that procedure one or (better) several years are left out, and we act as though they never occurred. The statistical model is developed on the retained years, then applied to the withheld year(s). This is done exhaustively for each year withheld. In some strategies one needs more than one level (i.e. nested) of cross-validation. In this way even a statistical tool may consume a lot of CPU. The science of cross-validation itself has to be further developed; several downsides and boobytraps have been noted (Barnston and Van den Dool 1993).

A forecast like the one in Figure 9.3 is thus based on both real-time aspects (*F*, the strength of the predictor) and hindcast aspects (*R*, the a priori correlation). Figure 9.6 shows a rendition of the real-time forecast by the CCA tool for the entire USA for MAM 2006. (Other tools are presented in the same way to the forecaster.) Here, in a nutshell, both

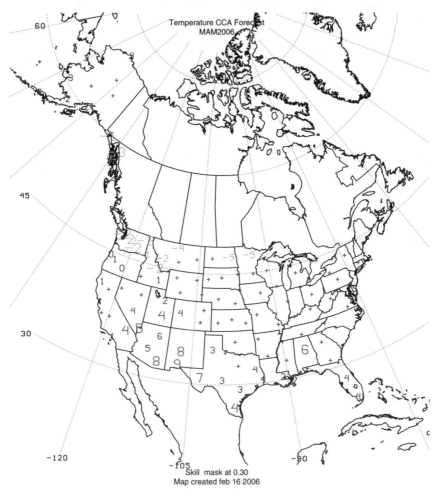

Figure 9.6 (see Plate 12) The looks of a tool used to make the seasonal prediction. The tool is the CCA. The units are standard deviations multiplied by 10. Red (blue) values are positive (negative) anomalies. The size of the numeral indicates the level of a priori skill. At the locations, indicated by only the red or blue plus sign, the a priori skill is below the 0.3 correlation, and no forecast value is shown.

the current state, and the hindcasts over 1955–last year play a role. The point forecast (in units of standard deviation × 10) is strongly non-zero if (a) CCA had, locally, skill over the past 50 years and (b) the predictor (modes of global SST mainly) is sufficiently anomalous. The cutoff for local skill is taken to be 0.3; any correlation below 0.3 is considered indistinguishable from zero, practically or statistically.

9.10   Verification 2: - Heidke skill scores

There is no substitute for making forecasts in real time and doing verification a posteriori. Here we report on a verification of the official (OFF) seasonal temperature forecast over 1995–2002, all 102 climate divisions, all 13 leads and all 12 initial seasons combined. The measure used, mainly for the sake of tradition, is the Heidke skill score, a categorical score generically defined as:

$$SS = (H - E)/(T - E) \times 100 \qquad (9.2)$$

where T = total number of forecasts, $E = T/3$ is the expected number of hits by chance, and H = number of categorical hits. Because we make EC forecasts it makes sense to verify the non-EC forecasts (the colored areas in Fig 9.4) first as:

$$SS_1 = (H_1 - E_1)/(T_1 - E_1) \times 100 \qquad (9.2a)$$

where T_1 is the total number of non-EC forecasts, $E_1 = T_1/3$, and H_1 the number correct. Coverage is defined as T_1/T. A score for the whole nation, SS_2, including the blanks (the EC area), is produced by counting EC forecast as 1/3 correct. One simply finds: $SS_2 = SS_1 \times$ coverage.

How do we judge a Heidke score of 20–25%? For readers more familiar with correlations: On a large set of forecasts, and modest skill, the Heidke score for a three-class system equals half the correlation (Barnston 1991), e.g. SS=20 corresponds to a 40% correlation. Relative to a probability forecast, one can convert by considering that on average the observed class has been forecast as the favored class 13–17% more often than expected by chance, i.e. the average probability shift (for $SS_1 = 20$) is comparable to what is shown in the example in Figure 9.3.

In Table 9.2 we only show tools that are used all the time, and have been archived from the beginning of the long lead prediction in late 1994. In order to compare the performance of two tools one is advised to compare SS_2. We thus conclude from Table 9.2 that the official forecast is better than the participating tools. This is mainly from increased coverage. Apparently CCA and OCN, while having similar SS_1, have skill in non-overlapping areas and the forecaster is capable of combining these two tools to arrive at

Table 9.2. Heidke skill score of CPC temperature seasonal forecasts for JFM95–FMA2002. All 102 climate divisions, starting times and leads are combined. The CCA and OCN methods were unchanged during this period, while the dynamical method (predecessor of current CFS) changed several times

	SS_1	SS_2	Coverage (%)
OFF	22.7	9.4	41.4 (13 leads)
CCA	25.1	6.4	25.5
OCN	22.2	8.3	37.4
Dynamical	7.6	2.5	32.7 (First four leads only)

a superior official forecast. (Keep in mind that the forecaster, on occasion, uses other tools, such as ENSO composites which were highly successful in winter 1997/98, see Barnston et al. 1999.) SS_1 is important to verify that the a priori skill estimates used in real time (but based on the historical record up to that time) were correct. Considering that we use a 0.3 correlation as cut-off one wants SS_1 to be in excess of 15. CCA and OCN's a priori skill estimates appear to be correct and holding up on independent data. The dynamical model used over 1995–2002 may not have been optimal. The a priori skill estimate was inaccurate, or the model in real time was not exactly the same as the one used for the hindcasts. We believe this has improved since summer 2004 when the CFS was introduced (Saha et al. 2006). On the whole CPC makes non-EC forecasts for about 40% of the nation. While this may seem a downer for the remaining 60%, one needs to keep in mind that it boosts faith in the non-EC forecast where and when they are issued. Forcing forecasters to make a non-EC forecast under all circumstances, especially when skill is certified low, is counterproductive.

A table like Table 9.2 for precipitation is not shown because all SS values are very low, between 0 and 5, dangerously close to a random forecast. An analysis as to which tool contributes the most seems meaningless. If it were not for an occasional strong ENSO winter the skill of precipitation might indeed be very close to zero. An analysis of OCN over the years 1962–present shows that OCN skill for precipitation in the 1960s through 1980s was generally better than it has been since 1995. Conversely, OCN skill for temperature was dangerously low in the 1980s when a regime of generally cold temperatures (all seasons in the 1960s and 1970s) was replaced by generally warmer temperatures after 1995, first in winter and to lesser extent in summer.

Much more detailed regional verification is forthcoming, see Halpert and Pelman (2004).

What Table 9.2 does not convey is that nearly all skill in temperature is due to shifting probabilities to above normal for temperature. This point is addressed in the next section.

9.11 Trends

We have mentioned low-frequency variations as a tool in making seasonal forecasts, especially for temperature, see OCN in Section 8.3. Table 9.2 shows that OCN has a strong contribution to the skill of the official forecast. The presence of trends has also a strong influence on the operational forecast in several other ways. Consider these facts: averaged across the United States temperatures for the 102 climate divisions over 1991–2005 have averaged 1.3 1.2 0.6 0.2 0.4 0.1 0.2 0.4 0.5 0.2 0.1 1.0°C above the 1961–1990 mean for the months January, February, March,...through December, respectively. When expressed as (local) standardized monthly data before taking the national mean these shifts are 0.5 0.5 0.3 0.2 0.4 0.1 0.2 0.4 0.4 0.1 0.0 0.4, respectively, for the 12 months.[12] Had we known this in advance (we did not!), probability shifts on the order of what is shown in Figure 9.3, which was just an educational example, could have been expected for virtual all seasons just based on "trends". (The skill of OCN suggests persistence of 10-year averaged anomalies as a workable tool in real time.) In some areas, like the South-West USA the shifts are stronger, while they are weaker in the northern plains, see Figure 8.2. One nowadays needs very strong inter-annual indications for a cold outcome in order to even dare to favor B. Keep in mind that many users round off the probability forecasts to a categorical (the one suggested by the color). Given how infrequent the B class is observed, for instance 7% instead of 33.3% of the time in 2005, see Table 9.3, forecasters nowadays shy away from placing high odds in the B class.

Table 9.3 shows that during 1995–1997 the observed frequency was not very different from expected. But from 1998 onward (and in spite of the earliest possible update to 1971–2000 normals in May 2001) the outcome has been predominantly A.

Of the four options, favoring either B, N, A, or EC (climatological probabilities), only A and EC are used frequently, thereby calling the three-class system into question, and forecast maps tend to look alike, regardless of lead, and to a lesser extent, regardless of season. It is only at the subtler level of probabilities that one can see the inter-annual component (due to ENSO or soil moisture) reduce the odds for above normal temperature that would be suggested by the trend alone. Managing this situation is a challenge, and an unannounced occasional cold month (with noteworthy societal impact) comes across as a huge bust. Given that in the 1960s and 1970s the trend was for persistent cold (relative to 30-year normals in effect at that time), see Gilman (1986), the forecasters today

---

[12] Introducing new normals in May 2001 has hardly lessened the bias. The difference of the 2001–2005 average relative to updated normals 1971–2000 is 1.0 −0.1 0.0 0.7 0.2 0.2 0.6 0.4 0.6 0.4 1.2 1.0°C for the 12 months.

Table 9.3. The observed frequency (%) of occurrence of the three terciles in seasonal mean temperatures across the USA

B	N	A	at 102 US locations*
26	28	46	1995
36	34	30	1996
27	32	41	1997
08	17	75	1998
13	24	63	1999
22	20	58	2000
15	32	53	2001 (Normals changed! To 1971–2000)
19	36	46	2002
15	38	47	2003
20	33	47	2004
07	32	61	2005

*assumed to be 1/3, 1/3, 1/3, based on 30-year 1961–1990 normals period.

wonder when the current warming trend is going to turn around. Often, so far erroneously, they feel it could be "now".

9.12 Forecasts of opportunity and the tension with regularly scheduled operations

The level of skill reported in this book is not terribly high. Moreover, practioners know that a modest overall 0.35–0.50 correlation often hides a simple truth, namely that on a few occasions we have some truly usable forecast skill and in the rest of the circumstances we have virtually no skill at all. Following this point of view into the extreme we should perhaps refrain from issuing forecasts, except when the opportunity looks good, for instance when there is a strong ENSO event coming. The idea of "forecasts of opportunity" is certainly not new, but is somewhat at odds with a regularly scheduled official forecast. Once a new forecast is expected each month it is hard to say: No, not now. The audience may no longer be there, when five years from now we finally see an opportunity. The way to manage this is by probabilities and in particular by using EC without being ashamed, and to make non-EC forecasts when and where the opportunity exists. In practice, however, there is considerable pressure to make non-EC forecasts more often, if not all of the time. For instance NOAA organizes press releases and conferences in the spring and fall as part of its annual activities calendar. A coast-to-coast EC forecast may not strike the audience as a great contribution to a newsworthy event, so the temptation is to put something on the map. This practicality has to be balanced against a more academic stand about forecast skill, credibility, and when and where we go for high probabilities.

Appendix: Historical notes

If someone wanted to describe how short-term climate prediction in the USA is done in practice over the years, the literature would be helpful but quite limited. This chapter describes the nuts and bolts of the seasonal forecast at CPC over the last five or 10 years, while Chapter 8 includes the methods used at CPC formally during this period. One may have to go back to Wagner (1989) for a previous review with some detail. Some of the motivation about going to long-lead forecasts in 1994 is given in a trio of workshop papers (Barnston 1994; Van den Dool 1994, O'Lenic 1994). From Gilman (1985; 1986) and Epstein (1988) one may surmise how the forecast in the US was made about 20 years ago. Gilmans's predecessor, Namias, was a prolific writer, and the period of the 1950s, monthly forecasts mainly, has been described quite extensively (Namias 1953). (Via Roads (1986) one can access more history about the Namias era, including Namias' collected works.) Van den Dool and Gilman (2004) summarized the influence of 50 years of NWP on the monthly/seasonal forecast. Finally there is a booklet for the 25th anniversary of CPC (Reeves and Gemmill 2004) with personal accounts by many of the forecasters and an attempt to write a formal history (Reeves et al. 2005).

10 Conclusion

In this book we have reviewed empirical methods in short-term climate prediction. We devoted a whole chapter to the design of two of these methods, Empirical Wave Propagation (EWP, Chapter 3) and Constructed Analogue (CA Chapter 7). Other methods of empirical prediction were listed in Chapter 8, with brief descriptions and examples and references. One chapter is devoted to EOFs, as such a diagnostic topic, but widely used in both prediction and diagnostics, and thoroughly debated for a few decades. Two brief chapters, written in support of the subsequent chapter, Teleconnections (Chapter 4), should make the discussion on EOFs more interesting, and the topic of effective degrees of freedom (Chapter 6) is indispensable when one wants to understand why and when natural analogues would work (or not), or how an analogue is constructed, or how any method using truncation works.

Most chapters can be read largely in isolation, but connections can be made of course between chapters. EWP is claimed to be useful, if not essential, in understanding teleconnections. Dispersion experiments, featuring day-by-day time-scales, link the CA and EWP methods. Examples of El Nino boreal winter behavior can be found in (a) the examples of EOFs on global SST and 500 mb streamfunction (Chapter 5), (b) specification of surface weather from 500 mb streamfunction (Chapter 7), and (c) the ENSO correlation and compositing approach (Chapter 8). The noble pursuit of knowledge may have been as important in the choice of some material as any immediate prediction application. Chapter 9 is different, less research oriented, and more an eyewitness description of what goes on in the making of a seasonal prediction. This eyewitness account style spills over into Chapter 8 here and there, because in order to understand why certain methods have survived to this day some practicalities have to be understood.

The closeness to real-time prediction throughout the book creates a sense of application. However, the application in this book does not go beyond the making of the forecast itself; we completely shied away from such topics as a cost/benefit analysis or decision-making process by, for example, a climate sensitive potato farmer or reservoir operator. Hartmann et al. (2002) describe the CPC forecasts with an eye towards users.

Some special topics are left for this, the final, chapter to be emphasized, organized, discussed, maybe solved, or left for the interested reader to pursue further. In this order we will discuss (1) linearity, (2) why GCMs do not (yet handily) outperform empirical methods, (3) predictability and (4) the future of short-term climate prediction. The linearity is discussed first because it is very important for the second topic.

10.1 Linearity

Although the equations of motion are nonlinear, some aspects of behavior in the atmosphere are perhaps much more linear than expected. To avoid confusion let us mention that the words linear and linearity are used in somewhat different ways in various contexts.

(a) The equations being nonlinear means that they contains terms in which products of basic variables occur, the most obvious example being a momentum equation

$$\frac{\partial u}{\partial t} = -u \frac{\partial u}{\partial x}, \text{ etc.} \tag{10.1}$$

where u is a wind component. In this case (non)linearity relates directly to time tendency on the left-hand side.

(b) The response of the atmosphere to El Nino and La Nina is said, by some, to be linear if the response to positive and negative SST anomalies is, except for the sign of the atmospheric anomaly, the same. This feature actually has more to do with the symmetry of the pdf with respect to the mean. Whether there is a relationship between type (a) dynamical and type (b) statistical linearity, we do not know, but see Burgers and Stephenson (1999) for a discussion.

(c) A linear operator. Given the equation $\mathbf{A}y = \mathbf{x}$, where $\mathbf{A}$ and $\mathbf{x}$ are known and y is solved for, the operator is said to be linear if $\mathbf{A}(\alpha y) = (\alpha \mathbf{x})$ for any value of α. Type (b) linearity follows from type (c). On occasion (Hoskins and Karoly 1981; Opsteegh and Van den Dool 1980), the atmosphere has been described as a linear steady state operator to describe teleconnections as a response to forcing. Doubling the forcing x, leads to doubling the response y.

(d) An important feature of linearity in some contexts is that orthogonal modes do not exchange energy.

Obviously no one stops us from linearizing Equation (10.1), to the extent possible, i.e. substitute some mean state (U) plus a departure from it (u'). There are no definitive rules to choose U, so we leave in the middle exactly what U is (climatology, the mean state in the absence of transients, today's state in all its details, etc.). The right hand side of (10.1) can be written, $u\partial u/\partial x = U\partial u'/\partial x + u'\partial U/\partial x + U\partial U/\partial x + u'\partial u'/\partial x$. Assuming there is an equation for the mean state, $\partial U/\partial t = -U\partial U/\partial x + \ldots + $ known terms, we can thus write

$$\frac{\partial u'}{\partial t} = -U\partial u'/\partial x - u'\frac{\partial U}{\partial x} - u'\frac{\partial u'}{\partial x}. \tag{10.2}$$

This recasts the forecast problem as one concerned with anomalies only (when U is climatology), a very familiar theme in this book. It is very possible that among the three terms on the right hand side $u'\partial u'/\partial x$ is not dominant. The prototypical nonlinear advection term can thus often be linearized to a considerable extent.

Instances in this book where linearity was noteworthy and possibly surprising are the following.

(1) EWP (Chapter 3) appears to yield a linear wave dispersion relationship as if each wave, in an aggregate sense, can be treated without regard for the presence of other waves. In other areas of study it has been difficult to find averaging procedures that make the nonlinearity small or vanishing. For instance, the mean state of the atmosphere can never be understood without including the mean effect of the transients (Peixoto and Oort 1992). The apparent linearity of EWP is related somewhat to using only short time separations in judging the phase speed. Presumably this is similar to the validity of tangent linear models for a duration of at most a few days. If nonlinearity dominated in (10.2) (i.e. $u'\partial u'/\partial x \gg U\partial u'/\partial x + u'\partial U/\partial x$) none of this would be possible.[1]

(2) CA (Chapter 7) appears to duplicate much of EWP when applied to the same set of one-day forecasts. The trick of CA is to linearly combine states, and their subsequent evolution. But what exactly happens to the equations when two (only two to keep it simple) previously observed states u'_1 and u'_2 are averaged: $u'_3 = (u'_1 + u'_2)/2$. One must wonder whether the average of the observed $\partial u'_1/\partial t$ and $\partial u'_2/\partial t$ is a good forecast for $\partial u'_3/\partial t$. Again the success of CA, modest as it is, would be impossible unless the third term in Equation (10.2) is small, or at least not dominant. See the appendix in Chapter 7 for more details.

(3) The success of CA in specification problems (Section 7.3; no time derivative involved) is evidence of quasi-symmetric pdfs. It may be odd to say that rain and sunshine are each other's opposite, but when thought of as the pdf of a variable like vertical motion, there is indeed near symmetry relative to a climatology. Similarly, the search for natural analogues and anti-analogues (Section 7.1) indicated only a very

---

[1] In the less restrictive application of phase shifting (Cai and Van den Dool 1991) we found that sub-harmonics of the wave one follows survive the averaging, i.e. EWP is not entirely linear.

small difference in their quality, suggesting linearity of type (b). Not shown is that the tendencies following (natural) anti-analogues, with sign reversed, are only slighty worse as a forecast than the tendencies following natural analogues (Van den Dool 1991), i.e. nonlinearity of type (a) cannot be dominant.

(4) There are plenty of linear correlations reported in this book. That the measurement of pressure at Darwin in Australia in SON (or the SST in a small area called Nino34) relates linearly to seasonal mean conditions half a world away for the next JFM should not cease to amaze us. How different is the situation in short-range forecasting. In order to make a 5-day forecast we need a global model and accurate initial data over the whole globe. In fact one missing observation has sometimes been blamed for a failed forecast. How can things be so simple, type (c) linearity, for the seasonal forecast?

(5) The linearity question has been raised with nearly all tools in Chapter 8, and addressed best in the literature in connection with the LIM. Why should Equation (8.8) $dx/dt = \mathbf{L}x + \mathbf{R}$ be approximately valid? The answer offered by Winkler et al. (2001) is that by taking a suitable (time) mean, like a weekly mean, the dynamical short time-scale non-linearities of type (a) can be represented as a stochastic process acting upon the more slowly and linearly evolving time mean state. Whitaker and Sardeshmukh (1998) even go as far as showing that the effect of storm tracks is linear in terms of anomalies in the time mean flow.

Consideration of linearity drives the answer in the next section.

10.2   Relative performance GCMs and empirical methods

In 2005 NCEP gave some of its employees an award for developing a global land–ocean–atmosphere system called the Climate Forecast System. The citation included a sentence that stated that for the first time in history numerical seasonal predictions were on par with empirical methods. From a competing institution we got a publication entitled: Did the ECMWF seasonal forecast model outperform a statistical model over the last 15 years? (Van Oldenborgh et al. 2005). How does one explain that after years of development, a costly GCM, absorbing enormous resources, does not handily outperform some simple empirical method. Why is it, in this day and age, that simple models, empirical methods, etc. are still of some value, and used at CPC and CDC in real-time seasonal forecast operations? This may in fact be unanticipated, after the optimism expressed some 15 years ago about using models for this purpose (Palmer and Anderson 1994). The answer in our opinion has to do with linearity. Most empirical methods are linear or very close to being linear. The only fundamental advantage a GCM has over linear methods is that it can execute the nonlinear terms. However, if for some magical reason the problem of seasonal forecasting is quite linear, a GCM cannot exploit its only advantage. Indeed following what we said in Section 10.1, this may already be the case in the week-2 forecast (Winkler et al. 2001) where LIM and the NCEP global spectral model are not very different in skill. On may see in the same light the definition of the practical limit of predictability in Saha and Van den Dool (1988). This limit is said to be reached if the continued model integration of an n-day forecast out to day $n+1$, is no better than persistence of the n-day forecast.

However, we do not need to go as far as requiring that the atmosphere is almost linear. No matter how small, nonlinearity is never trivial.[2] Instead, we argue below that as far as forecast skill in short-term climate prediction is concerned atmospheric (and oceanic) models are functionally linear. The lines of reasoning are as follows:

(1) Empirical methods are linear, or nearly so.[3]
(2) Physical models have one clear advantage over empirical models: they can execute the nonlinear terms.
(3) A model needs at least three degrees of freedom to be nonlinear (Lorenz 1960) so as to allow energy exchange among modes, although not any three d.o.f. will do.
(4) We speculate that a nonlinear model with nominally millions of degrees of freedom, but skill in only ≤ 3 d.o.f. is functionally linear in terms of the skill of its forecasts, and, to its detriment, the nonlinear terms add random numbers to the time tendencies of the modes with prediction skill. It takes large ensembles to remove this noise.
(5) Empirical methods like analogues, see Section 7.1, given ~ 50 years of data, can deal very well with about three effective degrees of freedom.

Therefore: Physical models need to have skill in, effectively, more than three d.o.f. before there is a scientific basis for expecting them to outperform linear empirical methods in a forecast setting.[4] The number three arises for two unrelated reasons. One needs at least three dofs for a nonlinear model, and given ~ 50 years of observations empirical methods should be very good at problems with three or less edofs. (It follows incidentally that the NA method, while methodically nonlinear, is also functionally linear as long as we cannot match more than three dofs.)

How many degrees of freedom (with forecast skill) do GCMs have in the seasonal forecast problem? It would appear that currently the answer is

---

[2] Here we are quoting H. Tennekes, who would say the same thing but dissipation instead of nonlinearity.
[3] Some efforts to make nonlinear empirical methods notwithstanding. Neural nets, and analogues are nonlinear in principle.
[4] We emphasize forecast setting here; dynamical models may be far better than empirical methods in a simulation mode, but not in prediction mode.

only around one, maybe two. Quan and Hoerling (2005) have shown that the lion's share of the GCM forecast skill can be duplicated by a linear regression on the first mode of tropical Pacific SST variability. That points to order one d.o.f. Anderson et al. (1999) have shown that models, vintage late 1990s, could not outperform CCA on identically the same task. Several authors (Straus and Shukla 2000) have found that the ensemble mean state of the model atmosphere in long AMIP runs has a very dominant first EOF, or, in our terms a very low N. That kind of simplicity is actually an argument against the application of GCMs to the forecast problem. That only one mode survives in the ensemble mean indicates that other modes, which do exist in individual members and in Nature, do not correlate among members, and disappear upon taking the ensemble mean, a sure sign of low predictability. The forecast of a single EOF can be done very well by empirical means; no models needed. In fact Anderson (1999) found that CCA trained on model data could make an equal or better forecast of the next member in the ensemble than a model integration itself. Even under perfect model assumption the dofs with forecast skill thus appear very limited.[5]

Given that only ENSO, trends and soil moisture come to mind as factors in short-term climate prediction, it is hard to imagine that the d.o.f. that can be skillfully predicted is very high.

A lack of nonlinearity in the seasonal forecast problem can also be seen in the benefit of systematic error correction. Figure 10.1 shows along the y-axis the anomaly correlation of bias corrected ensemble mean Z500 forecasts[6] for AMIP runs by several models (over the period 1950–1994, see Peng et al. (2002) for details), as a function of the magnitude of the systematic error in Z500. GCMs have substantial systematic error. The standard deviation for seasonal mean Z500 in the NH is around 30 gpm in DJF, so the four models shown in Figure 10.1 have a systematic error ranging from one up to three times the natural variability. If the models were operating in a highly nonlinear environment the simulation of anomalies would be much less (more) skillful for models with the largest (smallest) systematic error. However, that dependence is only weakly present. Apparently even a huge systematic error can be removed and the anomalies salvaged, a sign of linearity which is unwelcome if GCMs are at some point in the future expected to exploit their ability to do nonlinear calculations and beat linear methods. Extrapolation of the four entries in Figure 10.1 to zero systematic error may be dangerous, but does not point to anything higher than a 0.5 to 0.6 anomaly correlation for a perfect model.

---

[5] Characterizing the d.o.f. that two data sets (like forecasts and verification) have in common needs more work than what was presented in Chapter 6.

[6] These are not strictly forecasts, since perfect SST is provided.

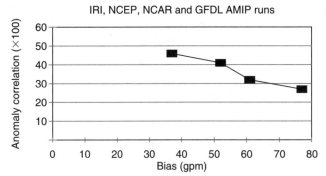

Figure 10.1 The anomaly correlation of bias corrected seasonal mean ensemble mean DJF Z500 simulations (in four AMIP simulations 1950–1994) as a function of the bias. Domain: 20°N–pole. The participating models are from IRI, NCEP, NCAR and GFDL.

Simply expecting that models will eventually handily outperform empirical methods, because this also happened in the short range, makes no sense. One may reason along the lines presented above, that in the short-range weather forecast, GCMs, which have skill in very many dofs initially, should be much better than any linear (dynamical or empirical) method. This has indeed been found to be the case since about 1965. We have to identify the dofs we may be able to predict a season ahead of time under ideal circumstances, then rationally proceed by deciding which tools are the best approach in a real-time forecast setting. This does not reduce the importance of models. A good simulation of atmosphere–ocean behavior on all time-scales is very important and has many applications. For instance, it is unlikely we can estimate short-term climate predictability from anything but dynamical models.

At this point the reasoning along the lines of effective degrees of freedom and functional linearity is conjecture. We have obviously not proven that the above lines of reasoning, steps (1) to (5), are correct.

10.3 Predictability

Predictability is thought of as the prediction skill one could achieve under ideal circumstances. Predictability is a ceiling for prediction skill. It helps to know predictability, so as to stay realistic, or to see how much improvement still awaits us. Below we give four approaches to determine prediction skill and predictability that have been in vogue over the last 40 years. Hopefully this inspires the readers to some original work where it is needed most.

Approach 1: Evaluation of skill of real-time prediction the old-fashioned way. Simply make forecasts and wait 25 or 50 years to see how well a method performs. Problems include (a) small sample size, and (b) a long waiting time (and funding agents are impatient).

Approach 2: Evaluation of skill of hindcasts. This removes the long waiting time aspect. Problems include (a) small sample size, (b) "honesty" of hindcasts (overfit, tuning problems and fundamental problems with cross-validation) and (c) hindcasts cannot be done for official forecasts, only strictly objective unambiguous methods.

Approach 3: Predictability of the first kind (sensitivity due to uncertainty in initial conditions); or, how long do two perturbed members in an ensemble stay closer together than randomly selected historical states? This method (Lorenz 1982) is very famous because of its clear connection to popular chaos theory. A model vs model verification amounts to a perfect model assumption. Problems include the choice of the size of the initial error and the nature of the error (growing, decaying...). Other problems arise if the spread of the ensemble members is low, or equivalently, the model's variability is lower than in nature. Studies of the predictability of the first kind have lead to the insight that day-by-day weather in the mid-latitudes is predictable for at most one or two weeks, depending on criterion.

Approach 4: Predictability of the second kind due to variations in "external" boundary conditions. This approach has come about mainly in the AMIP context (prescribed global SST variations) and goes by the names potential predictability, reproducibility, etc. Problems include unclarity about the lead of the forecast the predictability pertains to, and what is meant by external. Madden's (1976) approach based on data (the only empirical method to estimate predictability we are aware of) also fits in this category.

AMIP runs have indicated extremely high predictability of the second kind in the tropics ($>50\%$ explained variance), but more modest or sobering numbers in mid-latitude.

Approach 1 is out, because of impatience. With the development of one-tier coupled atmosphere–ocean models, it appears that approach 4 has come to an end also. Only sensitivity to initial conditions (including erstwhile boundary conditions like SST) survives.[7] This leaves us only with evaluating hindcasts and predictability of the first kind as major tasks. Neither task is trivial.

There has been no developmental work after about 1985 on the method suggested by Madden (1976) who gave empirically based estimates of ~25% potential predictability in mid-latitudes for seasonal mean surface

---

[7] Only CO_2 increase, solar variability, atmospheric turbidity and the like come to mind as surviving external factors.

pressure. Except for CA (this book) and NA (Eshel 2006) empirical methods damp anomalies to zero as skill goes down with lead, and are thus unsuitable for predictability estimates. One needs diverging solutions to study predictability.

The definition of predictability, even the first kind, may need to be reworked. The older approaches were generally based on traditional verification, such as rms, anomaly correlation, etc. But with the advent of creating full membered model ensembles new approaches using pdfs directly may have to be invented. Some recent work by DelSole (2004) already points in this direction.

Finally, the perfect model assumption needs work. While members of an ensemble obey exactly the same physics and numerics, why should the predictability estimate apply to the real world? In the end one must demand models to be good replicas of nature, including faithful simulation of, say, the Madden and Julian Oscillation and the QBO. The MJO is thought to be important for the shorter lead climate forecasts, but predictability estimates cannot be taken seriously when the models don't have an MJO, or have a weak MJO with erroneous phase speed.

We recently evaluated predictability of the first kind using the CFS model (Saha et al. 2006), which is a state of the art one-tier global ocean–atmosphere model. A single member was correlated against the average of 14 remaining members. The news is mixed. On the downside, the predictability of T and P over land in the NH does not exceed 0.4 correlation for any lead/verifying month. A more positive note is that the erstwhile boundary conditions (w and SST) are not only predicted well but have even higher predictability. Over the oceans, modeling has apparently advanced to the point where we can beat the control forecast (persistence) for SST everywhere (not just in the equatorial Pacific). Over land, the high skill in w forecasts is not as high as skill of persistence forecasts of w. This is indicative of problems with forecasting the sum of P minus evaporation minus runoff.

One must also understand that estimates of predictability look better, by definition, when ensemble members have low spread, low compared to the rmse of the control forecast. Other researchers appear to interpret low spread as motivation to add stochastic forcing to the models, so the increased spread and rmse are the same (and prediction skill and predictability become identically the same and all hope for improvement is gone).

10.4   The future of short-term climate prediction

We end this book on prediction with speculation about the future. What will the state of short-term climate prediction be in the future? And what are the priorities? A conservative approach to looking into the future is to extrapolate the advances of the last 10 years. Among the main advances:

(1) The number of models or methods available in real time is increasing rapidly. Prior to 2000 organizations like CPC had only a handful of in-house tools. Technology and fast communication have allowed outsiders access to data they need to run their methods and make output available in a timely fashion. In spite of sobering estimates of prediction skill (and predictability) in mid-latitudes, the enthusiasm is enormous, both among modelers and empiricists. With so many new forecasts (hundreds of them) consolidation (Chapter 8) is an increasing priority.

(2) Massive hindcast data sets. Each method should be accompanied by a hindcast data set, and increasingly this is what is happening. While the shining examples may have been for the seasonal prediction (Palmer et al. 2004; Saha et al. 2006), the generation of hindcasts is spilling over to the shorter forecast ranges as well. In order to run hindcasts, the reanalysis of land, atmosphere and ocean (as far back as possible) is an increasing priority. Empirical studies are a major beneficiary of reanalyses also.

(3) Probabilistic approach from beginning (perturbed IC) to end (application models). Although the seasonal prediction was one of the first to be expressed probabilistically, there has only recently been a major push towards probability expression and verification. To a certain extent this may have been a spill-over from the experience in ensemble forecasting in the medium range which gave new life to probability forecasting. Serious users are well served by reliable probability forecasts. The casual user may notice less benefit and feel excluded by the high abstraction level. Exactly how pdfs will be constructed from ~100 or 1000 members, each with a weight (based on skill and possibly collinearity) remains a subject of study.

Two more topics where advances are required are the following.

(4) We need to come to grips with long-term trends. Although trend tools, like OCN, are an ingredient in the seasonal prediction there is a dearth of methods contributing anything original about inter-decadal variability. The emphasis has been on the inter-annual time-scale and ENSO. But now that the occurrence of the below normal temperature tercile has become a rare event we may need to adjust methods and presentation. A connection with longer term climate change research may be a natural avenue of progress with mutual benefits.

(5) Since we work under a cloud of low predictability, we need to agree on how to define predictability, develop the notion, understand caveats and develop minimum requirements for any model to be used in a perfect model setting for a quasi-definitive predictability estimate.

References

Achatz, U. and Opsteegh, J. D. (2003). Primitive-equation-based low-order models with seasonal cycle. Part I: Model construction. *J. Atmos. Sci.*, **60**, 465–477.

Ambaum, M. H. P., Hoskins, B. J., and Stephenson, D. B. (2002). 'Arctic oscillation or North Atlantic oscillation', *J. Climate*, **15**, 1969–1978.

Anderson, J. L. (1991). The robustness of barotropic unstable modes in a zonally varying atmosphere. *J. Atmos. Sci.*, **48**, 2393–2410.

Anderson, J. L., van den Dool, H. Barnston, A. Chen, W. Stern W. and Ploshay, J. (1999). Present-day capabilities of numerical and statistical models for atmospheric extratropical seasonal simulation and prediction. *Bulletin of the American Meteorological Society*, **80**, 1349–1361.

Anderson, J. L., (1999). Why are statistical models for seasonal prediction competitive with current generation GCM predictions? In *Proceedings of the 24th Annual Climate Diagnostics and Prediction Workshop*, Springfield, VA: NTIS, 176–178.

Baer, F., and Platzman, G. W. (1961). A procedure for numerical integration of the spectral vorticity equation. *J. Atmos. Sci.*, **18**, 393–401.

Baer, F. (1972). An alternate scale representation of atmospheric energy spectra. *J. Atmos. Sci.*, **29**, 649–664.

Barnett, T. P. and Preisendorfer, R. W., (1978). Multifield analogue prediction of short-term climate fluctuations using a climate state vector, *J. Atmos. Sci.*, **35**, 1771–1787.

Barnett, T. P., (1981). Statistical prediction of North American air temperature from Pacific predictors, *Monthly Weather Review*, **9** (5), 1021–1041.

Barnett, T. P. and Preisendorfer, R. (1987). Origins and levels of monthly and seasonal forecast skill for United States surface air temperatures determined by canonical correlation analysis. *Monthly Weather Review*, **115**, 1825–1850.

Barnston, A. G., and Livezey, R. E. (1987). Classification, seasonality and persistence of low frequency atmospheric circulation patterns, *Monthly Weather Review*, **115**, 1083–1126.

Barnston, A. G. (1992). Correspondence among the correlation, RMSE, and Heidke forecast verification measures; refinement of the Heidke score. *Weather and Forecasting*, **7**, 699–709.

Barnston, A. G. and van den Dool, H. M. (1993). A degeneracy in cross-validated skill in regression based forecasts. *J. Climate*, **6**, 963–977.

Barnston, A. G., (1994). Linear statistical short-term climate predictive skill in the northern hemisphere. *J. Climate*, **7**, 1513–1564.

Barnston, A. G., Chelliah, M. and Goldenberg, S. B. (1997). Documentation of a highly ENSO-related SST region in the equatorial Pacific: Research Note. *Atmosphere-Ocean* 1997 **35**(3): 367–383.

Barnston, A. G., A. Leetmaa, V. E. Kousky, R. E. Livezey, E. O'Lenic, H. Van den Dool, A. J. Wagner and D. A. Unger. (1999). NCEP Forecasts of the El Niño of 1997–98 and its US Impacts. *Bulletin of the American Meteorological Society*, **80**, 1829–1852.

Barnston, A. G., S. J. Mason, L. Goddard, D. G. Dewitt and S. E. Zebiak. (2003). Multimodel ensembling in seasonal climate forecasting at IRI. *Bulletin of the American Meteorological Society*, **84**, 1783–1796.

Barnston, A. G., and Smith, T. M. (1996). Specification and prediction of global surface temperature and precipitation from global SST using CCA. *J. Climate*, **9**, 2660–2697.

Barry, R. G., and Carleton, A. M. (2001). *Synoptic and Dynamic Climatology*. Routledge, London.

Berlage, H. P. (1957). Fluctuations of the general atmospheric circulation of more than one year, their nature and prognostic value. Kon. Ned. Meteorol. Inst. Mededelingen en Verhandelingen. **69**,

Black, R. X., McDaniel, B. A. and Robinson, W. A. (2005). Stratosphere–troposphere coupling during spring onset *J. Climate* (submitted).

Branstator, G. (1987). A striking example of the atmosphere's leading traveling pattern. *J. Atmos. Sci.*, **44**, 2310–2323.

Branstator, G. (1995). Organization of storm track anomalies by recurring low-frequency circulation anomalies. *J. Atmos. Sci.*, **52**, 207–226.

Branstator, G. and J. Berner. (2005). Linear and nonlinear signatures in the planetary wave dynamics of an AGCM: Phase space tendencies. *J. Atmos. Sci.*, **62**, 1792–1811.

Bretherton, C.S., M. Widmann, V. P. Dymnikov, J. M. Wallace and I. Bladé. (1999). The effective number of spatial degrees of freedom of a time-varying field. *J. Climate*, **12**, 1990–2009.

Burgers, G. and Stephenson, D. B. (1999). The Normality of El Nino, *Geophys. Res. Letters*, **26**, 1027–1030.

Cai, M. and van den Dool. H. M. (1991). Low-frequency waves and traveling storm tracks. Part I: Barotropic component. *J. Atmos. Sci.*, **48**, 1420–1436.

Cai, M. and Van Den Dool. H. M. (1992). Frequency waves and traveling storm tracks. Part II: Three-dimensional structure. *J. Atmos. Sci.*, **49**, 2506–2524.

Cai, M. and Van Den Dool. H. M. (1994). Dynamical decomposition of low-frequency tendencies. *J. Atmos. Sci.*, **51**, 2086–2100.

Cayan, D. R., K. T. Redmond and L. G. Riddle. (1999). ENSO and hydrologic extremes in the Western United States. *J. Climate*, **12**, 2881–2893.

Chang, E., and I. Orlanski. (1994). On energy flux and group velocity of waves in baroclinic flows. *J. Atmos. Sci.*, **51**, 3823–3828.

Chang, F-C., and J. M. Wallace. (1987). Meteorological conditions during heat waves and droughts in the United States Great Plains. *Monthly Weather Review*, **115**, 1253–1269.

Chelliah, M. and G. D. Bell. (2004). Tropical multidecadal and interannual climate variability in the NCEP–NCAR reanalysis. *J. Climate*, **17**, 1777–1803.

Chen, W.Y., (1982). Assessment of Southern Oscillation sea-level pressure indices. *Monthly Weather Review*, **110**, 800–807.

Chandrasekaran, S., and Schubert, K.E. (2005). Models for robust estimation and identification. Internet link: http://ftp.csci.csusb.edu/ schubert/pubs/SurveyICCSA03.pdf

Chang, J. J. C., and Mak, M. (1993). A dynamical empirical orthogonal function analysis of the intraseasonal disturbances. *J. Atmos. Sci*, **50**, 613–630.

Compo, G. P., and P. D. Sardeshmukh. (2004). Storm track predictability on seasonal and decadal scales. *J. Climate*, **17**, 3701–37.

Court, A., (1967–68). Climate normals as predictors: Parts I–V. Science Reports, Air Force Cambridge Research Laboratory Bedford, MA, Contract AF19(628)–5176.

DelSole, T., (2004). Predictability and information theory. Part I: Measures of predictability. *J. Atmos Sci.*, **61**, 2425–2440.

Dickson, R. R., (1967). The climatological relationship between temperatures of successive months in the United States. *J. Applied Meteorology*, **6**, 31–38.

Douglas, A. V., and P. J. Englehart. (1981). On a statistical relationship between autumn rainfall in the central equatorial Pacific and subsequent winter precipitation in Florida. *Monthly Weather Review*, **109**, 2377–2382.

Douglas, A. V., D. R. Cayan and J. Namias. (1982). Large-scale changes in North Pacific and North American weather patterns in recent decades. *Monthly Weather Review*, 110, 1851–1862.

Draper, N.R. and Smith, H. (1981). *Applied Regression Analysis*. Wiley Series in Probability and Mathematical Statistics. John wiley, New York.

Eliasen, E, and Machenhauer, B, A study of the fluctuations of the atmospheric planetary flow patterns represented by spherical harmonics, *Tellus*, 17, May 1965, 220–238.

Epstein, E. S. (1988). Long-range weather prediction: Limits of predictability and beyond. *Weather and Forecasting*, 3, 69–75.

Epstein, E. S. (1991). On obtaining daily climatological values from monthly means. *J. Climate*, 4, 365–368.

Eshel, G., (2006). Empirically evaluating divergence rates of atmospheric trajectories *J. Atmos. Sci.*, 63(2), 741–753.

Fan, Y., H. Van den Dool, K. Mitchell, D. Lohmann, (2006). 1948–1998 US hydrological reanalysis by the Noah Land Assimilation System. *J. Climate*. 19, 1214–1237.

Farrell, B. (1984). Modal and non-modal baroclinic waves. *J. Atmos. Sci.*, 41, 668–673.

Fraedrich, K., (1986). Estimating the dimensions of weather and climate attractors. *J. Atmos. Sci.*, 43, 419–432.

Fraedrich, K., C. Ziehmann and F. Sielmann. (1995). Estimates of spatial degrees of freedom. *J. Climate*, 8, 361–369.

Franzke, C., Y. Lee and S. B. Feldstein. (2004). Is the North Atlantic Oscillation a breaking wave? *J. Atmos. Sci.*, 61, 145–160.

Frederiksen, J. S. and G. Branstator, (2001). Seasonal and intraseasonal variability of large-scale barotropic modes. *J. Atmos. Sci.*, 58, 50–69.

Frederiksen, J. S. and G. Branstator. (2005). Seasonal variability of teleconnection patterns. *J. Atmos. Sci.*, 62, 1346–1365.

Gandin, L. S. (1965). *Objective Analysis of Meteorological Fields*, translated by Isr. *Prog. Sci. Transl.*, 242 pp., Gidrometeoizdat, Leningrad, Russia.

Gates, W. L., J. S. Boyle, C. Covey, C. G. Dease, C. M. Doutriaux, R. S. Drach, M. Fiorino, P. J. Gleckler, J. J. Hnilo, S. M. Marlais, T. J. Phillips, G. L. Potter, B. D. Santer, K. R. Sperber, K. E. Taylor and D. N. Williams. (1999). An Overview of the Results of the Atmospheric Model Intercomparison Project (AMIP I). *Bulletin of the American Meteorological Society*, 80, 29–55.

GEWEX, (2004). GEWEX Continental-Scale International Project (GCIP)-3. Forty articles reprinted from *J. Geophysical Research*. Special GCIP/GAPP volume,

Ghil M., Allen, R. M. Dettinger, M. D. Ide, K. Kondrashov, D. Mann, M. E. Robertson, A. Saunders, A. Tian, Y. Varadi, F. and Yiou, P. (2002) (PDF File). "Advanced spectral methods for climatic time series," *Rev. Geophys.*, 40(1), 3.1–3.41,

Gilman, D. L., Fuglister F.J. and Mitchell Jr. J.M. (1963). On the Power Spectrum of "Red Noise". *J. Atmos. Sci.*, 20, 182–184.

Gilman, D. L. (1985). Long-Range Forecasting: The Present and the Future. *Bulletin of the American Meteorological Society*, 66, 159–164.

Gilman, D. L. (1986). Expressing Uncertainty in Long Range Forecasts. *Proceedings of Namias symposium*, La Jolla, California. Edited by J. Roads. 174–187.

Gilchrist, A.,(1986). Long-range Forecasting. Q.J.R.M.S., 112, 567–592.

Glahn, H. R., and Lowry, D. A. (1972). The use of Model Output Statistics (MOS) in objective weather forecasting. *J. Appl. Meteor.*, 11, 1203–1211.

Gray, W. M., Landsea, C. W. Mielke, P. W., Jr., and Berry, K.J. (1994a). Predicting Atlantic basin seasonal tropical cyclone activity by 1 June. *Weather and Forecasting*, 9, 103–115.

Green, P. J. and Silverman, B. W. (1994). Nonparametric Regression and Generalized Linear Models. Chapman and Hall. London.

Hagedorn, R., Doblas-Reyes, F. J. and Palmer, T. N. (2005). The rationale behind the success of multi-model ensembles in seasonal forecasting. Part I: Basic concepts. *Tellus*, 57, 234–257.

Halpert, M. S., and Pelman, K. (2004). Regional verification of CPC's seasonal forecasts. 29th Annual Climate Diagnostics and Prediction Workshop, October 18–22, 2004 in Madison, WI. See http://www.cpc.ncep.noaa. gov/products/outreach/ proceedings/cdw29_proceedings/ presentations.shtml

Halpert, M. S., and C. F. Ropelewski. (1992). Surface Temperature Patterns Associated with the Southern Oscillation. *J. Climate*, 5, 577–593.

Hamill, T. M., J. S. Whitaker and X. Wei. (2004). Ensemble Reforecasting: Improving Medium-Range Forecast Skill Using Retrospective Forecasts. *Monthly Weather Review*, 132, 1434–1447.

Hamill, T. M., Whitaker, J. S. and Mullen, S. L. (2006). Reforecasts, an important new data set for improving weather predictions. *Bulletin of the American Meteorological Society.*, 87, 33–46.

Hansen, B. K., (2000). Analog forecasting of ceiling and visibility using fuzzy sets, 2nd Conference on Artificial Intelligence, American Meteorological Society, 1–7.

Hartmann, H. C., T. C. Pagano, Sorooshian, S. and Bales, R. (2002). Confidence Builders: Evaluating Seasonal Climate Forecasts from User Perspectives. *Bulletin of the American Meteorological Society*, 83, 683–698.

Hartmann, D. L. (1995). A PV View of Zonal Flow Vacillation. *J. Atmos. Sci.*, 52, 2561–2576.

Hastenrath, H. and L. Greischar. (1993). Further Work on the Prediction of Northeast Brazil Rainfall Anomalies. *J. Climate*, 6, 743–758.

Hastenrath, S., (2003). Climate Prediction (Empirical and Numerical). P411–417 in Encyclopedia of Atmospheric Sciences, Six-Volume Set, editors James R. Holton, Judy A. Curry, and John A. Pyle.

He, Y. X., and Barnston, A. G. (1996). Long-lead forecasts of seasonal precipitation in the tropical Pacific islands using CCA. *J. Climate*, 9, 2020–2035.

Higgins, R. W., Kim, H.-K. and Unger, D. (2004). Long-Lead Seasonal Temperature and Precipitation Prediction Using Tropical Pacific SST Consolidation Forecasts. *J. Climate*, 17, 3398–3414.

Hoerling, M. P., A. Kumar and M. Zhong. (1997). El Niño, La Niña, and the Nonlinearity of Their Teleconnections. *J. Climate*, 10, 1769–1786.

Holton, J. R. and Lindzen, R. S. (1972). An updated theory for the Quasi-Biennial cycle of the tropical stratosphere. *J. Atmos. Sci.* 29, 1076–1080.

Holton, J. R., (1979). *An Introduction to Dynamic Meteorology.* 2nd edition. Int. Geophysical Series, 23 Academic Press, New York.

Horel, J. D. (1981). 'A rotated principal component analysis of the interannual variability of the Northern Hemisphere 500 mb height field.', *Monthly Weather Review*, 109, 813–829.

Horel, J. D., and J. M. Wallace. (1981). Planetary-Scale Atmospheric Phenomena Associated with the Southern Oscillation. *Monthly Weather Review*, 109, 813–829.

Hovmoller, E., (1949). The Trough and Ridge Diagram. *Tellus*, 1, 62–66.

Huang, J., and H. M. van den Dool. (1993). Monthly Precipitation-Temperature Relations and Temperature Prediction over the United States. *J. Climate*, 6, 1111–1132.

Huang, J., H. M., van den Dool, and K. P. Georgarakos. (1996). Analysis of Model-Calculated Soil Moisture over the United States (1931–1993) and Applications to Long-Range Temperature Forecasts. *J. Climate*, 9, 1350–1362.

Hurrell, J. W. (1995). 'Decadal trends in the North Atlantic Oscillation', *Science*, 269, 676–679.

Hoskins, B. J., and Karoly, D. J. (1981). The Steady Linear Response of a Spherical Atmosphere to Thermal and Orographic Forcing. *J. Atmos. Sci.*, 38, 1179–1196.

Huang, J., H. M. van den Dool and A. G. Barnston. (1996). Long-Lead Seasonal Temperature Prediction Using Optimal Climate Normals. *J. Climate*, **9**, 809–817.

Hwang, S. O., Schemm, J. K. E., Barnston, A. G., and Kwon, W. T. (2001). Long-lead seasonal forecast skill in far eastern Asia using canonical correlation analysis. *J. Climate*, **14**, 3005–3016.

Johansson, Å, A. Barnston, S. Saha and H. van den Dool. (1998). On the Level and Origin of Seasonal Forecast Skill in Northern Europe. *J. Atmos. Sci.*, **55**, 103–127.

Johansson, Å, (2006). Prediction Skill of the NAO and PNA from Daily to Seasonal Time-scales. *J. Climate*, **17**., under review.

Jolliffe, I. T. (2002). Principal Component Analysis. Springer, New York, 2nd edition.

Jolliffe, I. T., and Stephenson, D. B. (Eds) (2003). Forecast Verification: A Practioner's Guide in Atmospheric Sciences. John Wiley New York, 240pp.

Kalnay, E., Kanamitsu, M., Kistler, R., Collins, W., Deaven, D., Gandin, L., Iredell, M., Saha, S., White, G., Woollen, J., Zhu, Y., Chelliah, M., Ebisuzaki, W., Higgins, W., Janowiak, J., Mo, K.C., Ropelewski, C., Wang, J., Leetmaa, A., Reynolds, R., Roy Jenne, and Dennis Joseph. (1996). The NMC/NCAR 40-Year Reanalysis Project". *Bulletin of the American Meteorological Society.*, **77**, 437–471.

Kharin, V., and F. V., Zwiers, W. (2002). Climate Predictions with Multimodel Ensembles. *J. Climate*, **15**, 793–799.

Kistler, R., Kalnay, E., Collins, W., Saha, S., White, G., Woollen, J., Chelliah, M., Ebisuzaki, W., Kanamitsu, M., Kousky, V., van den Dool, H., Jenne, R., Fiorino, M. (2001). The NCEP–NCAR 50–Year Reanalysis: Monthly Means CD–ROM and Documentation. *Bulletin of the American Meteorological Society*, **82**, 247–268.

Klein, W. H., and John, E., Walsh. (1983). A Comparison Of Pointwise Screening and Empirical Orthogonal Functions in Specifying Monthly Surface Temperature from 700 mb Data. *Monthly Weather Review*, **111**, 669–673.

Klein, W. H., and Hal Bloom, J. (1987). Specification of Monthly Precipitation over the United States from the Surrounding 700 mb Height Field. *Monthly Weather Review*, **115**, 2118–2132.

Klein, W. H., (1985). Space and Time Variations in Specifying Monthly Mean Surface Temperature from the 700 mb Height Field. *Monthly Weather Review*, **113**, 277–290.

Knaff, J. A, and Landsea, C.W. (1997). An El Nino–Southern Oscillation CLImatology and PERsistence (CLIPER) Forecasting Scheme. Weather *Forecasting*, **12**, 633–652.

Kravtsov, S., Kondrashov, D., and Ghil, M. (2005). Multilevel Regression Modeling of Nonlinear Processes: Derivation and Applications to Climatic Variability. *J. Climate*, **18**, 4404–4424.

Kruizinga, S., and Murphy, A. (1983). Use of an analogue procedure to formulate objectively probabilistic temperature forecasts in The Netherlands. *Monthly Weather Review*, **111**, 2244–2254.

Kushnir, Y. (1987). Retrograding wintertime low-frequency disturbances over the North Pacific Ocean. *J. Atmos. Sci.*, **44**, 2727–2742.

Kutzbach, J. E. (1967). Empirical Eigenvectors of Sea-Level Pressure, Surface Temperature and Precipitation Complexes over North America. *J. Applied Meteorology*, **6**, 791–802.

Landsea, C. W., Knaff, J.A. (2000). "How much skill was there in forecasting the very strong 1997–98 El Nino?" *Bulletin of the American Meteorological Society*, **81**, 2107–2119.

Lanzante, J. R. (1990). The Leading Modes of 10–30 Day Variability in the Extratropics of the Northern Hemisphere during the Cold Season. *J. Atmos. Sci.*, **47**, 2115–2140.

Lamb, P. J., and Stanley, A., Changnon Jr. (1981). On the "best" temperature and precipitation normals: The Illinois situation. *J. Applied Meteorology*, **20**, 1383–1390.

Lean, J., and David Rind. (1999). Evaluating sun–climate relationships since the Little Ice Age *J. Atmospheric and Solar–Terrestrial Physics*, **61** (1999) 25–36.

Lin, H., and J. Derome. (2004). Nonlinearity of the Extratropical Response to Tropical Forcing. *J. Climate*, **17**, 2597–2608.

Lindgrén, S., and Neumann, J. (1980). Great Historical Events That Were Significantly Affected by the Weather: 5, Some Meteorological Events of the Crimean War and Their Consequences. *Bulletin of the American Meteorological Society*, **61**, 1570–1583.

Livezey, R. E., and Chen, W. Y. (1983). Statistical field significance and its determination by Monte Carlo techniques. *Monthly Weather Review.*, **111**, 49–59.

Livezey, R. E., and Mo K. C. (1987). Tropical–extratropical teleconnections during the Northern Hemisphere winter. Part II: Relationships between monthly mean Northern Hemisphere circulation patterns and proxies for tropical convection. *Monthly Weather Review.*, **115**, 3115–3132.

Lorenz E. (1956). Empirical orthogonal functions and statistical weather prediction. Scientific report no 1., Air Force Cambridge Research Center, Air Research and Development Command, Cambridge, Mass.

Lorenz, E. N. (1960). Maximum simplification of the Dynamical Equations. Tellus, **12**, 243–254.

Lorenz, E. N. (1963). Deterministic Nonperiodic Flow. *J. Atmos. Sci.*, **20**, 130–141.

Lorenz, E. N. (1969). Atmospheric predictability by naturally occurring analogs. *J. Atmos. Sci.*, **26**, 636–646.

Lorenz, E. N. (1973). On the Existence of Extended Range Predictability. *J. Applied Meteorology*, **12**, 543–546.

Lorenz, E. N. (1982). Atmospheric predictability experiments with a large numerical model. *Tellus* **34**, 505–513.

Madden, R. A., and P. R., Julian. (1971). Detection of a 40–50 Day Oscillation in the Zonal Wind in the Tropical Pacific. *J. Atmos. Sci.*, **28**, 702–708.

Madden, R. A. (1978). Further Evidence of Traveling Planetary Waves. *J. Atmos. Sci.*, **35**, 1605–1618.

Madden, R. A. (1976). Estimates of the Natural Variability of Time-Averaged Sea-Level Pressure. *Monthly Weather Review*, **104**, 942–952.

Madden, R. A. (1989). On Predicting Probability Distributions of Time-Averaged Meteorological Data. *J. Climate*, **2**, 922–925.

Mason, S. J., and Goddard, L. (2001). Probabilistic precipitation anomalies associated with ENSO. *Bulletin of the American Meteorological Society*, **82**, 619–638.

McGabe, G. J., Palecki, M., and Betancourt, J. L. (2004). Pacific and Atlantic Ocean influences on multidecadal drought frequency in the United States. *Proc. NAS*, **101**, 4136–4141.

Michaelsen, J. (1987). Cross-Validation in Statistical Climate Forecast Models. *J. Applied Meteorology*, **26**, 1589–1600.

Miyakoda, K., Hembree, G. D., Strickler, R. F., and Shulman, I. (1972). Cumulative Results of Extended Forecast Experiments I. Model Performance for Winter Cases. *Monthly Weather Review*, **100**, 836–855.

Mo. K. C. (2003). Ensemble Canonical Correlation Prediction of Surface Temperature over the United States. *J. Climate*, **16**, 1665–1683.

Mooley, D. A., Parthasarathy, B., and Pant, G. B. (1986). Relationship between Indian Summer Monsoon Rainfall and Location of the Ridge at the 500-mb Level along 75°E. *J. Applied Meteorology*, **25**, 633–640.

Murphy, A. H., and E. S., Epstein. (1989). Skill Scores and Correlation Coefficients in Model Verification. *Monthly Weather Review*, **117**, 572–582.

Namias, J. (1952). The annual course of month-to-month persistence in climatic anomalies. *Bulletin of the American Meteorological Society*, **33**, 279–285.

Namias, J. (1953). Thirt-day forecasting: a review of a ten-year experiment. *Meteor. Monographs, Amer. Meteor. Soc.*, **2**, #6, 83.

Namias, J. (1981). Teleconnections of 700mb height anomalies for the Northern Hemisphere. California Cooperative Oceanic Fisheries Investigations (Calcofi) Atlas#29, Scripps Institution of Oceanography, La Jolla, CA.

Newman, M., and P. D., Sardeshmukh. (1995). A Caveat Concerning Singular Value Decomposition. *J. Climate*, **8**, 352–360.

Nicolis, C. (1998). Atmospheric Analogs and Recurrence Time Statistics: Toward a Dynamical Formulation. *J. Atmos. Sci.*, **55**, 465–475.

North, G. R., Bell, T.L., Cahalan, R.F., and Moeng, F.J. (1982). Sampling errors in the estimation of empirical orthogonal functions. *Monthly Weather Review.*, **110**, 699–706

O'Lenic, E., A., and R. E., Livezey. (1988). Practical Considerations in the Use of Rotated Principal Component Analysis (RPCA)in Diagnostic Studies of Upper-Air Height Fields. *Monthly Weather Review*, **116**, 1682–1689.

O'Lenic, E., A. (1994). A new paradigm for production and dissemination of the NWS' Long Lead seasonal climate outlooks. *Proceedings of the 19th Annual Climate Diagnostics Workshop*, November 14–18, 1994, College parke, MD, 408–411.

O'Lenic, E. A., and S. Handel. (2004). Extended-range analogue ensemble forecasts. In *Climate Diagnostics and Prediction Workshop*, Madison Wisconsin. See link: http://www.cpc.ncep.noaa.gov/products/outreach/proceedings/cdw29_proceedings/presentations.shtml Poster 4.17

O'Connor, J. (1969). Hemispheric teleconnections of mean circulation anomalies at 700 millibars. ESSA Tech Report WB10, U.S. Department of Commerce, Silver Spring, MD.

Opsteegh, J. D., and Van Den Dool, H. M. (1980). Seasonal Differences in the Stationary Response of a Linearized Primitive Equation Model: Prospects for Long-Range Weather Forecasting?. *J. Atmos. Sci.*, **37**, 2169–2185.

Palmer, T. N., Anderson, D.L.T. (1994). The prospect for seasonal forecasting - a review paper. *Q.J. R. Meteorol. Soc.*, **120**, 755–793.

Panofsky, H. A., and Brier, G. W. (1968). Some applications of Statistics to Meteorology. Pennsylvania State University, p224.

Patil, D. J., Hunt, B. R., Kalnay, E., Yorke, J. A., and Ott, E. (2001). Local Low Dimensionality of Atmospheric Dynamics, *Phys. Rev. Lett*, **86**, (2001), #26, 5878–5881.

Peng, P., and van den Dool, H. (2002). Ajusting the OCN prediction methods by invoking EOFs. *27th Climate Prediction and Diagnostics Workshop*. George Mason University, October, 21–25, 2002. See http://www.cpc.ncep.noaa.gov/products/outreach/proceedings/cdw27_proceedings/index.html

Peng, P., Kumar, A., van den Dool, H., and Barnston, A.G. (2002), An analysis of multimodel ensemble predictions for seasonal climate anomalies, *J. Geophys. Res.*, **107**(D23), 4710.

Penland, C., and Th. Magorian. (1993). Prediction of Niño 3 Sea Surface Temperatures Using Linear Inverse Modeling. *J. Climate*, **6**, 1067–1076.

Peixoto, J. P., and Oort, A. H. (1992). Physics of Climate. American Institute of Physics.

Plumb, R. A. (1977). The Interaction of two internal waves with the mean flow: implications for the theory of the quasi-biennial oscillation. *J. Atmos. Sci.*, **34**, 1847–1858.

Preisendorfer, R. W.(Postumously Edited by Curtis Mobley), (1988). Principal Component Analysis in Meterology and Oceanography (Developments in Atmospheric Science, 17) Elsevier Amsterdam.

Portis, D. H., J. E. Walsh, M. El Hamly and P. Lamb, J. (2001). Seasonality of the North Atlantic Oscillation. *J. Climate*, **14**, 2069–2078.

Qin, J., and van den Dool, H. M. (1996). Simple extensions of an NWP model. *Monthly Weather Review.*, **124**, 277–287.

Quadrelli, R., Bretherton, C. S., and Wallace, J. M. (2005). On sampling errors in empirical orthogonal functions, *J. Climate*, **18**, 3704–3710.

Quan, X., Hoerling, M., Whitaker, J., Bates, G., and Xu, T. (2006). Diagnosing Sources of U.S. Seasonal Forecast Skill, *J. Climate*, **19**, 3279–3293.

Reed, R. G., Campbell, W. J., Rasmussen, L. A., and Rogers, D. G. (1961). Evidence of downward-propagating annual wind reversal in the equatorial stratosphere. *J. Geophysical Research*, **66**, 813–818.

Reeves, R., and Gemmill, D. (2004). Climate Prediction Center. Reflections on 25 Years of Analysis, Diagnosis and Prediction, 1979–2004. US Government Printing Office.

Reeves, R., D. Gemmill, R., E., Livezey, J. Laver. (2005). Extending Atmospheric Forecasts Beyond Weather: The History of Climate Prediction. Proceedings of AMS annual meeting, San Diego, January 2005. Session 5.8.

Richman, M. B. (1986). Rotation of principle components. Int. *J. Climatology*, **6**, 293–335.

Richman, M. B., and Peter Lamb, J. (1985). Climatic Pattern Analysis of Three- and Seven-Day Summer Rainfall in the Central United States: Some Methodological Considerations and a Regionalization. *J. Applied Meteorology*, **24**, 1325–1343.

Roads, J. (1986). (Ed.): *Proceedings of Namias Symposium*, Scripps, University of California, La Jolla, California. P200

Rodwell, M. J., and Folland, C. K. (2002). Atlantic air-sea interaction and seasonal predictability. *Q.J. R. Met. Soc.*, **128**, 1413–1443.

Montroy, D. L., M. B., Richman, and P. Lamb, J. (1998). Observed Nonlinearities of Monthly Teleconnections between Tropical Pacific Sea Surface Temperature Anomalies and Central and Eastern North American Precipitation. *J. Climate*, **11**, 1812–1835.

Rinne, J., and Karhilla, V. (1975). A spectral barotropic model in horizontal empirical orthogonal functions. *Quart. J. Roy. Meteor. Soc.*, **101**, 365–382.

Roads, J., and Barnett, T. P. (1984). Forecasts of the 500 mb Height Using a Dynamically Oriented Statistical Model. *Monthly Weather Review*, **112**, 1354–1369.

Ropelewski, C. F., and Halpert, M.S. (1986). North American Precipitation and Temperature Patterns Associated with the El Niño/Southern Oscillation (ENSO). Monthly Weather Review, **114**, 2352–2362.

Ropelewski, C. F., and Halpert, M.S. (1987). Global and Regional Scale Precipitation Patterns Associated with the El Niño/Southern Oscillation. Monthly Weather Review, **115**, 1606–1626.

Ropelewski, C. F., and Halpert, M.S. (1989). Precipitation Patterns Associated with the High Index Phase of the Southern Oscillation. *J. Climate*, **2**, 268–284.

Rossby, C. G. (1939). 'Relation between Variations in the Intensity of the Zonal Circulation of the Atmosphere and the Displacement of the Semi-Permanent Centres of Action', *J. Marine Res.*, **2**, 38–55.

Rossby, C. G. (1941). The scientific basis of modern meteorology. Climate and Man, Yearbook of Agriculture 1941, 599–655.

Roulston, M. S., and Smith, L.A. (2003). Combining dynamical and statistical ensembles. *Tellus*, **55A**. 16–30.

Rukhovets, L. (1963). The optimum representation of the vertical distribution of certain meterorological elements. *Izv. Acad. Sci. USSR, Geophys. Ser.*, 626–636 (Russian), 391–396(English).

Saha, S., and van den Dool, H.M. (1988). A measure of the practical limit of predictability. *Monthly Weather Review.*, **116**, 2522–2526.

Saha, S., Nadiga, S., Thiaw, C., Wang, J., Wang, W., Zhang, Q., van den Dool, H. M., Pan, H.-L., Moorthi, S., Behringer, D., Stokes, D., White, G., Lord, S., Ebisuzaki, W., Peng, P., Xie, P. (2006). The NCEP Climate Forecast System. *J. Climate*, **19**, 3483–3517.

Schemm, J., van den Dool, H., Huang, J., and Saha, S. Construction of daily climatology based on the 17-year NCEP/NCAR Reanalysis. *Proceedings of 1st Reanalysis Workshop*. WMO/TD-No. 876, WCRP-104, 290–293, Silver Spring, MD, 27–31, Oct. 1997.

Schubert, S. D. (1985). A Statistical-Dynamical Study of Empirically Determined Modes of Atmospheric Variability. *J. Atmos. Sci.*, **42**, 3–17.

Schuurmans, C. J. E. (1973). Four-year experiment in long-range weather forecasting, using circulation analogues. *Meteor. Rundsch.*, **26**, 2–4.

Selten, F. M. (1993). Toward an Optimal Description of Atmospheric Flow. *J. Atmos. Sci.*, **50**, 861–877.

Shabbar, A, and Barnston, A. G. (1996). Skill of seasonal climate forecasts in Canada using canonical correlation analysis. *Monthly Weather Review.*, **124**, 2370–2385.

Shukla, J., Anderson, J., Baumhefner, D., Brankovic, C., Chang, Y., Kalnay, E., Marx, L., Palmer, T., Paolino, D., Ploshay, J., Schubert, S., Straus, D., Suarez, M., and Tribbia, J. (2000). Dynamical Seasonal Prediction. *Bulletin of the American Meteorological Society*, **81**, 2593–2606.

Simmons, A. J. and B. J., Hoskins. (1979). The Downstream and Upstream Development of Unstable Baroclinic Waves. *J. Atmos. Sci.*, **36**, 1239–1254.

Simmons, A., Wallace, J.M., and Branstator, G.W. (1983). Barotropic wave propagation and instability, and atmospheric teleconnection patterns. *J. Atmos. Sci.*, **40**, 2383–2398.

Salby, M. L. (1982). A Ubiquitous Wavenumber-5 Anomaly in the Southern Hemisphere During FGGE. *Monthly Weather Review*, **110**, 1712–1721.

Smith, T. M., and R. E., Livezey. (1999). GCM Systematic Error Correction and Specification of the Seasonal Mean Pacific—North America Region Atmosphere from Global SSTs. *J. Climate*, **12**, 273–288.

Smith, T. M., and R. W., Reynolds. (2004). Improved Extended Reconstruction of SST (1854–1997). *J. Climate*, **17**, 2466–2477.

Smith, T. M., and R. Reynolds, W. (2004). Reconstruction of Monthly Mean Oceanic Sea Level Pressure Based on COADS and Station Data (1854–1997). *J. Atmospheric and Oceanic Technology*, **21**, 1272–1282.

Stephenson, D.B. (1997). Correlation of spatial climate/weather maps and the advantages of using the Mahalanobis metric in predictions, *Tellus*, **49**A, 513–527.

Stephenson, D. B., Coelho, C.A.S., Balmaseda, M., and Doblas-Reyes, F.J. (2005). Forecast Assimilation: A unified framework for the combination of multi-model weather and climate predictions, *Tellus* **57**A, (DEMETER special issue).

Straus, D. M., and Shukla, J. (2000). Distinguishing between the SST-forced variability and internal variability in mid-latitudes: Analysis of observations and GCM simulations. *Quart. J. Roy. Met. Soc.*, **126**, 2323–2350.

Straus, D., and Shukla, J. (2002). Does ENSO Force the PNA? *J. Climate*, **15**, 2340–2358.

Unger, D., A. Barnston, H. van den Dool and V. Kousky. (1996) onward. Consolidated Forecasts of Tropical Pacific SST in Nino 3.4 Using Two Dynamical Models and Two Statistical Models. *Experimental Long Lead Forecast Bulletin.* 1996 onward.

Unger, D., : Forecasts of Surface Temperature and Precipitation Anomalies over the U.S. Using Screening Multiple Linear Regression. *Experimental Long Lead Forecast Bulletin.* 1996 onward.

Tikhonov, A., and Arsenin, V. Y. *Solutions of Ill-Posed Problems.* Winston, Washington, 1977.

Tippett, M., A. G. Barnston, D. G. DeWitt and R. -H. Zhang. (2005). Statistical Correction of Tropical Pacific Sea Surface Temperature Forecasts. *J. Climate*, **18**, 5141–5162.

Thiaw, W. M., Barnston, A. G., and Kumar, V. (1999). Predictions of African rainfall on the seasonal timescale. *J. Geophys. Res.*, (A), **104** (D24), 31589–31597.

Thompson, D. W. J., and Wallace, J. M. (1998). The Arctic Oscillation signature in winter-time geopotential height and temperature fields. *Geophys. Res. Lett.*, **25**, 1297–1300.

Thompson, P. Duncan, (1977). How to improve accuracy by combining independent forecasts. *Monthly Weather Review* **105**: 228–229.

Torrence, C., Compo, G. P. (1998). A Practical Guide to Wavelet Analysis. *Bulletin of the American Meterological Society*, **79**, 61–78.

Toth, Z. (1995). Degrees of freedom in Northern Hemisphere circulation data. *Tellus*, **47**A, 457–472.

Toth, Z., and E. Kalnay. (1997). Ensemble Forecasting at NCEP and the Breeding Method. *Monthly Weather Review*, 125, 3297–3319.

Tracton, M. S., and E. Kalnay. (1993). Operational Ensemble Prediction at the National Meteorological Center: Practical Aspects. *Weather and Forecasting*, 8, 379–398.

Trenberth, K. E. (1984). Some Effects of Finite Sample Size and Persistence on Meteorological Statistics.Part II: Potential Predictability. *Monthly Weather Review*, 112, 2369–2379.

Trenberth, K. E. (1990). Recent Observed Interdecadal Climate Changes in the Northern Hemisphere. *Bulletin of the American Meteorological Society*, 71, 988–993.

Van den Dool, H. M. (1981). An indirect estimate of the predictability of the monthly mean atmosphere. *Scientific Report, Royal Netherlands Meteorological Service*, 81–6.

Van den Dool, H. M. (1983). A Possible Explanation of the Observed Persistence of Monthly Mean Circulation Anomalies. *Monthly Weather Review*, 111, 539–544.

Van den Dool, H. M., and Nap, J.L. (1985). Short and Long Range Air Temperature Forecasts near an Ocean. *Monthly Weather Review*, 113, 878–887.

Van den Dool, H. M., Klein, W.H., and Walsh, J.E. (1986). The Geographical Distribution and Seasonality of Persistence in Monthly Mean Air Temperatures over the United States. *Monthly Weather Review*, 114, 546–560.

Van den Dool, H. M., and Chervin, R.W. (1986). A comparison of month-to-month persistence of anomalies in a general circulation model and in the Earth's atmosphere. *J. Atmos. Sci.*, 43, 1454–1466.

Van den Dool, H. M. (1989). A New Look at Weather Forecasting through Analogues. *Monthly Weather Review*, 117, 2230–2247.

Van den Dool, H. M., and Toth, Z. (1991). Why do forecasts for near-normal fail to succeed? *Weather and Forecasting*, 6, 76–85.

Van den Dool, H. M. (1994). Searching for analogues, how long must one wait? *Tellus*, 46A, 314–324.

Van den Dool, H. M. (1994). New operational long-lead seasonal climate outlooks out to one year: Rationale. *19th Climate Diagnostics Workshop*, Nov, 14–18, 1994. 405–407.

Van den Dool, H. M., and Barnston, A.G. (1994). Forecasts of Global Sea Surface Temperature out to a Year using the Constructed Analogue Method. *Proceedings of Climate Diagnostics Workshop*, 19, College Park, MD, November 14–18, 1994, pp.416–419.

Van den Dool, H. M., and Rukhovets, L. (1994). On the Weights for an Ensemble-Averaged 6–10-Day Forecast. *Weather and Forecasting*, 9, 457–465.

Van den Dool, H. M., and Qin, J. (1996). An efficient and accurate method for continuous time interpolation of large-scale atmospheric fields, *Monthly Weather Review*, 964–971.

Van den Dool, H. M., Saha, S., Schemm, J., and Huang, J. (1997). A temporal interpolation method to obtain hourly atmospheric surface pressure tides in renalysis 1979–95. *J. Geoph. Res*, 102, D18, 22013–22024.

Van den Dool, H. M., Saha, S., and Johansson, A. (2000). Empirical orthogonal teleconnections. *J. Climate*, 13, 1421–1435.

Van den Dool, H., Huang, J., and Fan, Y. (2003). Performance and analysis of the constructed analogue method applied to U.S. soil moisture over 1981–2001, *J. Geophys. Res.*, 108(D16), 8617.

Van Loon H., and Rodgers, J. C. (1978). The seasaw in winter temperatures between Greenland and Northen Europe. Part I: General description. *Monthly Weather Review.*, 106, 296–310.

Van Loon, H., and R. Madden, A. (1981). The Southern Oscillation. Part I: Global Associations with Pressure and Temperature in Northern Winter. *Monthly Weather Review*, 109, 1150–1162.

Van Oldenborgh, G. J., M. Balmaseda, A., L. Ferranti, Timothy Stockdale, N., and D. Anderson, L. T. (2005). Did the ECMWF Seasonal Forecast Model Outperform Statistical ENSO Forecast Models over the Last 15 Years? *J. Climate*, 18, 3240–3249.

Vautard, R., G. Plaut, R. Wang and G. Brunet. (1999). Seasonal Prediction of North American Surface Air Temperatures Using Space–Time Principal Components. *J. Climate*, **12**, 380–394.

von Storch, H., and Zwiers, F. W. (1999). Statistical Analysis in Climate Research. Cambridge University Press. Cambridge.

Wagner, A. J. (1989). Medium- and Long-Range Forecasting. *Weather and Forecasting*, **4**, 413–426.

Wagner, A. J., and Maisel, N. (1989). NH 700 mb Teleconnections Based on Half-Monthly mean Anomaly data stratified by Season and by sign. *Proceedings of the th Climate Diagnostics Workshop*, 228–229.

Wallace, J. M., and Gutzler, D.S. (1981). 'Teleconnections in the geopotential height field during the Northern Hemisphere winter', *Monthly Weather Review.*, **109**, 784–812.

Wallace, J. M. (2000). North Atlantic Oscillation / Annular Mode: two paradigms - one phenomenon. *Q. J. Royal Met. Soc.*, **126**, 791–805.

Waliser, D., C. Jones, J. -K. Schemm, E., and N. Graham, E. (1999). A Statistical Extended-Range Tropical Forecast Model Based on the Slow Evolution of the Madden–Julian Oscillation. *J. Climate*, **12**, 1918–1939.

Waliser, D., K. Weickmann, R. Dole, S. Schubert, O. Alves, C. Jones, M. Newman, H. -L. Pan, A. Roubicek, S. Saha, C. Smith, H. van den Dool, F. Vitart, M. Wheeler, J. Whitaker. (2006). The experimental MJO Prediction Project. *Bulletin of the American Meteorological Society*, **87**, 425–431.

Walker, G. T. (1924). Correlation of seasonal variations in weather IX: A further study of world weather. *Mem. Indian Meteor. Dep.*, **24**, 275–332.

Wang, X., and S. Shen, S. (1999). Estimation of Spatial Degrees of Freedom of a Climate Field. *J. Climate*, **12**, 1280–1291.

Washington, W., and Parkinson, C. (1986). An Introduction to Three-Dimensional Climate Modeling, Oxford University Press Oxford.

Webster, P. J. and C. Hoyos. (2004). Prediction of Monsoon Rainfall and River Discharge on 15–30-Day Time Scales. *Bulletin of the American Meteorological Society*, **85**, 1745–1765.

Whitaker, J. S., and Sardeshmukh, P.D. (1998). A linear theory of extratropical synoptic eddy statistics. *J. Atmos. Sci.*, **55**, 237–258.

White, G.H. (1980). Skewness, Kurtosis and Extreme Values of Northern Hemisphere Geopotential Heights. *Monthly Weather Review*, **108**, 1446–1455.

Wilks, D. (1995/2005). Statistical Methods in the Atmospheric Sciences : An Introduction (International Geophysics Series). ISBN: 0127519653. Updated as 2nd edition in November 2005.

Wilks, D. S. (1996). Statistical Significance of Long-Range "Optimal Climate Normal" Temperature and Precipitation Forecasts. *J. Climate*, **9**, 827–839.

Winkler, C. R., M. Newman and P. Sardeshmukh, D. (2001). A Linear Model of Wintertime Low-Frequency Variability. Part I: Formulation and Forecast Skill. *J. Climate*, **14**, 4474–4494.

Wolter, K., R. Dole, M., and C. Smith, A. (1999). Short-Term Climate Extremes over the Continental United States and ENSO. Part I: Seasonal Temperatures. *J. Climate*, **12**, 3255–3272.

Wu, A., W. Hsieh, and A. Shabbar. (2005). The Nonlinear Patterns of North American Winter Temperature and Precipitation Associated with ENSO. *J. Climate*, **18**, 1736–1752.

Wu, W., and Dickinson, RE. (2005). Warm-season rainfall variability over the US Great Plains and its correlation with evapotranspiration in a climate simulation. *Geophysical Research Letters*, **32**, L17402, doi:10.1029/2005GL023422.

Xue, Y., A. Leetmaa and M. Ji. (2000). ENSO Prediction with Markov Models: The Impact of Sea Level. *J. Climate*, **13**, 849–871.

Index

Appendices, figures, notes and tables are indexed in bold as **ap, f, n, t** e.g. **172f**